Advanced Sciences and Tec for Security Applications

C000247082

Series Editor

Anthony J. Masys, Associate Professor, Director of Global Disaster Management, Humanitarian Assistance and Homeland Security, University of South Florida, Tampa, USA

Advisory Editors

Gisela Bichler, California State University, San Bernardino, CA, USA

Thirimachos Bourlai, Lane Department of Computer Science and Electrical Engineering, Multispectral Imagery Lab (MILab), West Virginia University, Morgantown, WV, USA

Chris Johnson, University of Glasgow, Glasgow, UK

Panagiotis Karampelas, Hellenic Air Force Academy, Attica, Greece

Christian Leuprecht, Royal Military College of Canada, Kingston, ON, Canada

Edward C. Morse, University of California, Berkeley, CA, USA

David Skillicorn, Queen's University, Kingston, ON, Canada

Yoshiki Yamagata, National Institute for Environmental Studies, Tsukuba, Ibaraki, Japan

Indexed by SCOPUS

The series Advanced Sciences and Technologies for Security Applications comprises interdisciplinary research covering the theory, foundations and domain-specific topics pertaining to security. Publications within the series are peer-reviewed monographs and edited works in the areas of:

- biological and chemical threat recognition and detection (e.g., biosensors, aerosols, forensics)
- crisis and disaster management
- terrorism
- cyber security and secure information systems (e.g., encryption, optical and photonic systems)
- traditional and non-traditional security
- energy, food and resource security
- economic security and securitization (including associated infrastructures)
- transnational crime
- human security and health security
- social, political and psychological aspects of security
- recognition and identification (e.g., optical imaging, biometrics, authentication and verification)
- smart surveillance systems
- applications of theoretical frameworks and methodologies (e.g., grounded theory, complexity, network sciences, modelling and simulation)

Together, the high-quality contributions to this series provide a cross-disciplinary overview of forefront research endeavours aiming to make the world a safer place.

The editors encourage prospective authors to correspond with them in advance of submitting a manuscript. Submission of manuscripts should be made to the Editor-in-Chief or one of the Editors.

More information about this series at http://www.springer.com/series/5540

Ryan N. Burnette
Editor

Applied Biosecurity: Global Health, Biodefense, and Developing Technologies

 Springer

Editor
Ryan N. Burnette
Merrick & Company
Washington, D.C., USA

ISSN 1613-5113   ISSN 2363-9466   (electronic)
Advanced Sciences and Technologies for Security Applications
ISBN 978-3-030-69466-1   ISBN 978-3-030-69464-7   (eBook)
https://doi.org/10.1007/978-3-030-69464-7

This Springer imprint is published by the registered company Springer Nature Switzerland AG
The registered company address is: Gewerbestrasse 11, 6330 Cham, Switzerland

Preface

Long before sophisticated biocontainment laboratories were designed, or the modern concept of bioterrorism evolved, before biodefense found its way into the vernacular, and before gene-editing technologies unlocked the power of genetic manipulation, the term "biosecurity" was already in use by forward-thinking agriculturists as they protected their livestock from infectious disease. Today, biosecurity is a term that means different things to different industries yet continues to be applied in domains as they perpetually evolve. Societies are coming to a reckoning with the advances of many aspects of biology and medicine, both viewed as good and bad. Perhaps this is why many in the fields of biology recognize the need to protect the significant value in the multitude of elements that comprise our growing, vital economy of biological advances. We are all stakeholders in the fields of biology as we are all dependent on its virtues. The result is a genuine need to protect those assets that quite literally impact the lives of billions every day.

The purpose of this book is not to debate, or even truly refine, the definition of the term "biosecurity." Rather, the purpose of this book is to challenge practitioners in the fields of biology, security, governance, agriculture, human health, defense, and a host of others, *apply* the concepts of biosecurity in meaningful ways in an effort to secure the wealth of biological assets on which we are all dependent. It is in the application of these concepts that fields of practice have the opportunity to converge, the ability to grow toward something greater than any one field could achieve independently. At the time of writing this book, our species finds itself in the midst of the greatest biological event in generations: the COVID-19 global pandemic. The pandemic has laid bare our vulnerabilities in health systems, global commerce, the economy, ethics, and even our individual ideals of safety and security. Perhaps now is an ideal time to consider biosecurity in a broader concept, beyond the walls of a single laboratory or fence line of a farm, and work toward greater application of biosecurity in a concerted, multifaceted capacity. The authors freely admit that biosecurity is not the sole answer to solving the current or next epidemic or pandemic event. Yet when our societies are faced with such an event it becomes necessary to investigate fully every opportunity for improvement. Therefore, this book aims to focus on the applications of biosecurity in the areas of global health, defense, and developing technologies: three areas of growth, debate, and promise.

A unique feature of this book is a discussion on one of the foundational elements of biosecurity regardless of application: the threat assessment. Arguably for the first time, the authors provide a framework that provides for the integration of the upstream threat assessment and downstream risk assessment, effectively linking two domains of biorisk management. In addition to promoting cross-collaboration amongst biosafety and biosecurity professionals, this provides a framework with the intended result of an integrated biorisk management program. In the laboratory, so many elements of biosecurity have historically been relegated to the biosafety professional, a group with tremendous biorisk assessment prowess but less-armed with the trades of threat- and vulnerability-based tools. Conversely, institutional security programs stand to benefit from greater awareness of the nuances surrounding biorisk management that demonstrate biosecurity is a unique discipline. But even this is perhaps a narrow perspective of the broader applications presented in this book.

This book explores the idea that the principles of biosecurity can be applied to very large-scale settings and events, such as global health and pandemics. Here again, we see the convergence of ideals where health and security share a nexus. Our species is becoming more aware that threats such as infectious disease have the ability to alter our individual and collective sense of what it means to be secure. Again, the COVID-19 pandemic has made this clear. But the potential gain of considering our health as a species through the lens of a critical asset is both stronger health systems and greater security.

The topic of biodefense is more and more prevalent in our society as we have become unfortunately accustomed to acts of terrorism. Yet, biodefense programs are not only in place to deter or respond to malicious acts involving bioweapons but also fortifies systems to stand against naturally occurring infectious disease. This convergence of public health and defense is permeating regulations and international guidance frequently and is embracing many laboratory-level biosecurity concepts. Here, we review some of the major biodefense and threat reduction efforts occurring in the U.S. and elsewhere. Like the global health sector, there is a need to adapt laboratory-level biosecurity measures and apply them beyond the confines of the laboratory to reduce threats in the field, the farm, and the environment.

The discussion of biodefense is timely, given the fact that this book is being written during the global COVID-19 pandemic. In so many ways the world has changed and the pandemic has exposed cracks in many levels of our societies. Here, we investigate the role and applications of biosecurity through the lens of epidemic and pandemic events. Further, we provide research to suggest that the pandemic will have lasting impacts on how we perceive biosecurity through a convergence of expertise, implications for individual- and population-level health security, a reduced threshold for acts of bioterrorism, and the critical role cyberbiosecurity will play in protecting the growing bioeconomy.

Beyond the shadow of a pandemic, and in some cases in response to the pandemic, technological advances continue to grow exponentially. The past decade has seen gene editing and synthetic biology skyrocket to ubiquity in both traditional and unconventional biology laboratories. The duality of this observation holds both great promise for advances and great threats for malicious actors. We reexamine the dual-use research of concern model against the backdrop of these technologies and offer applications of biosecurity that reach beyond the typical biology laboratory. We provide an argument that unconventional research domains will require unconventional applications of biorisk management if we are to reap the reward of progress while simultaneously stifling attempts to use these technologies for illicit purposes.

The goal of this book is to provide biologists, security professionals, biorisk management experts, public health stakeholders, regulators, and the broad base of biology practitioners, as well as sympathizers, with a thought-provoking dialog to advance the discussion of the applications of biosecurity. Many texts have sought to capture the philosophy and principles of biosecurity: here, we attempt to further those bodies of expertise and provide frameworks to actually apply those principles. The authors sincerely hope this manuscript will provide a variety of practitioners more confidence in examining their respective operations, as they all contribute some value to the biological advances that cure, feed, and enrich us all.

Washington, D.C., USA
December 2020

Ryan N. Burnette, Ph.D.
Ryan.burnette@merrick.com

Acknowledgments

First and foremost, my sincere thanks to the contributors that helped complete this volume of what we all believe to be a topic of growing importance. It is a blessing to have professional relationships with such caring, passionate, experts who graciously donated their time to see this project to its conclusion. I would also like to thank Ms. Kelsie Judd for her editorial support and expert eye that dramatically raised the quality of this volume. Likewise, Ms. Samantha Dittrich, a corresponding author in this book, donated her time to ensure major sections of this work met expectations. Dr. Lauren Richardson would like to thank Dr. Stephen Goldsmith and Ms. Gabby Essix for their contributions. To my employer, Merrick & Company, I owe great gratitude. They allowed me time to organize, author, and edit this book. Further, they are the origin of many opportunities that will continue to see topics in this book manifest as actions. My sincere thanks also to Mr. Brad Andersen who has proven to be a mentor in many ways. Many of us that supported this book wish him the best on his next steps. To my wife, Kady, I always appreciate the patience and support I receive with projects such as these. And finally, to the readers of this book, allow all of those that contributed to this effort the opportunity to thank you for taking these matters into your own lives and professions—after all, it is left to all of us that consider ourselves stakeholders in the biosecurity domain to be the catalysts that advance these topics into practice toward the betterment of our world.

December 2020   Ryan N. Burnette, Ph.D.

Contents

About the Editor

Ryan N. Burnette, Ph.D. is an international practitioner of biorisk management, biocontainment laboratory operations consulting, and biodefense initiatives. He has worked with domestic agencies, foreign governments, academia, healthcare, industry, and independent laboratory programs in more than thirty countries. Dr. Burnette has published in the fields of molecular biology, endocrinology, biosafety, biosecurity, and infectious diseases and is the author and editor of one of the most recognized volumes in the field of biosecurity, "Biosecurity: Understanding, Assessing, and Preventing the Threat," published in 2013. Dr. Burnette has held positions in the departments of Biology and Biochemistry at Virginia Tech as well as the Department of Molecular Physiology and Biophysics at Vanderbilt University School of Medicine. He is currently a Vice President at Merrick & Company, a Colorado-based consulting corporation, where he oversees the Life Sciences practice from Merrick's branch office in the Washington, D.C. area.

Redefining Biosecurity by Application in Global Health, Biodefense, and Developing Technologies

Ryan N. Burnette

Abstract The term "biosecurity" has been broadly applied to a variety of industries over several decades, and continues to have various meanings to different audiences. It is generally accepted that the origin of the term can be traced back nearly one hundred years referencing certain agricultural practices in the context of controlling livestock health. Today, the term biosecurity can be widely seen in the laboratory environment, the media, in the context of biodefense, throughout government agencies and ministries worldwide, in discussions of genetically-modified organisms, agriculture, and information technology. Concerns of global health and security could not be more relevant than today, given the paradigm shifts manifested by the SARS-CoV-2 virus and the resulting coronavirus pandemic. This book will focus on the challenges that this and future pandemic events will bring and offer biosecurity as a set of practices to aid in humanity's combat of devastating emerging infectious diseases. To further add complexity to the definition are the rapid advances in technologies that challenge our understanding of how to apply biosecurity principles and practices. Finally, unlike the well-established field of biological safety (i.e., biosafety), there remains a lack of credentialing programs to define a "biosecurity professional." Taken together, our modern and technologically-driven world is in need redefining what biosecurity means and how it is practiced.

1 Historical Context of Biosecurity

In the strictest definition of biosecurity, the term expands to mean "security of biological assets." Such a definition casts a wide net of interpretation and usage. While this book will primarily focus on applications of biosecurity from the perspectives of laboratory programs, biodefense initiatives, and public health, it is worthwhile to understand the historical context of the term and the associated practices.

R. N. Burnette (✉)
Merrick & Company, Washington, D.C., USA
e-mail: ryan.burnette@merrick.com

© Springer Nature Switzerland AG 2021
R. N. Burnette (ed.), *Applied Biosecurity: Global Health, Biodefense, and Developing Technologies*, Advanced Sciences and Technologies for Security Applications,
https://doi.org/10.1007/978-3-030-69464-7_1

1

The use of biological agents as weapons (i.e., bioweapons) is not a new, or even a modern, concept. It was'nt until the 1990's that the U.S. launched regulatory measures to address issues related to bioweapons, dual-use research of concern, registering agents as "select," and the implementation of laws to regulate the ownerships, use, and transport of these select agents. In the U.S., this is codified under the Federal Select Agent Program (FSAP) (7 CFR Part 331; 9 CFR Part 121; 42 CFR Part 73). Internationally, the primary body of authority resides in the United Nations Biological Weapons Convention (BWC).

Today, the term biosecurity has so many contexts that it has become important to distinguish the sector when discussing biosecurity programs. Whether animal, plant, collections of cells, or other biological agents, the commonality of biosecurity across all spectrums is the need to protect the biological asset itself. As this book will detail in the following chapters, the foundation of any biosecurity program is the ability to identify and manage threats and vulnerabilities, thereby limiting the resulting risks. In other words, threat management becomes the fundamental element of any biosecurity program. For this reason, there is a second historical context of biosecurity- the history of threat assessment and management. This review requires an analysis of the law enforcement community and the tools that have been developed to prevent threats (usually people) from causing harm to assets (usually other people). Chapter two "The Biothreat Assessment as a Foundation for Biosecurity" will discuss in detail the employment of these theories and tools, and how they have been and can be adapted to biosecurity.

2   Distinguishing Biosecurity from Biosafety

Biosafety remains a well-established, growing discipline of laboratory practice, aimed at identifying and minimizing the risks presented by harmful biological agents to laboratorians, the public, animals, and the environment. Today, biosafety is best codified through several guidance documents that continue to be widely used, distributed, and used as a foundation for laboratory biosafety programs around the world.

In the U.S., the primary vehicle for biosafety guidance continues to be *Biosafety in Microbiological and Biomedical Laboratories* (BMBL), published jointly by the U.S. Centers for Disease Control and Prevention (CDC) and the National Institutes of Health (NIH), and recently released its sixth edition [3]. This set of guidelines has been widely adopted at both regulated and unregulated laboratory programs in the U.S.

The hallmark of biosafety as a field of practice is the process that defines risks and consequences of release of or exposure to potentially harmful biological agents. This process is known as the risk assessment, or in the context of biological agents, the *biorisk assessment*. In general, this process takes into account the properties of the biological agent (e.g., bacteria, viruses, fungi, toxins), forecasts the potential risks of handling, and establishes criteria and methods for minimizing or mitigating these risks. The result is a system of theory and practice that allows laboratorians to safely work with biological agents that have potentially dangerous properties. Of course the end goal of biosafety is the reduction of exposures and infections to laboratorians and others. In this context, the biological agent is the threat.

Counter this dogma with biosecurity. Biosecurity, on the other hand, is a field of practice that employs theory and methods to minimize the threats *to* the biological agents. In this context, the biological agents are in fact the assets to be protected. This brings forward the concept of directionality in the assessment process to be discussed in more detail below. The thrust of this book will be upon driving the application of the practices and principles of biosecurity as a distinct, yet complementary, field of practice to biosafety.

Unlike biosafety, the field of biosecurity is largely not codified in any single volume or collection. Further, the risk assessment process underpinning biosafety does not have a complementary threat assessment process specific to biological agents. It is worth noting, however, that guidance on biosecurity does comprise a section of the current BMBL [3] and the World Health Organization's *Laboratory Biosafety Manual* [13]. From a guidance perspective, this demonstrates a tangible overlap between biosafety and biosecurity concerns.

In the United States, and in many other countries, there are regulations with respect to the possession, handling, and transport of particularly dangerous biological agents. In the U.S. these are referred to as "select agents" and are regulated under the FSAP, jointly administered by CDC and the U.S. Department of Agriculture (USDA) under U.S. codes of federal regulation. While these codes provide mandatory requirements for working with biological agents of certain risk potentials, they do not offer prescriptive guidance toward establishing threat- or vulnerability-based criteria for biosecurity programs.

Also of note is that both ABSA (American Biological Safety Association) International and the International Federation of Biosafety Associations (IFBA) both offer varying types of biosafety credentialing programs for professionals in the field. These are experience- or experience and education-based credentials that signify a graduated comprehension and application of biosafety expertise. No credentialing program for biosecurity professionals currently exists, although ABSA International is currently evaluating the market and context for a biosecurity credentialing program.

3 Directionality of Biosecurity Matters

There is a distinct relationship between biosafety and biosecurity, and, as this book will impress upon, there is a need for significant overlap and interoperability to achieve comprehensive biorisk management. The continuity of biorisk- and biothreat-based tools and programs working synergistically is the foundation to protect the human, animal, and environmental resources *from* the biological assets (e.g., harmful bacteria and viruses) and to protect the biological assets themselves from threats imposed upon them. It is this paradigm of directionality that becomes critical to understanding the relationship between biosafety and biosecurity. In other words, depending on what point in the threat-to-asset-to-risk perspective you are, or the direction you are looking, reveals distinctions between threats and risks. For example, is a threat really a threat if it is not directionally moving toward an asset?

3.1   Directionality of the Biorisk Assessment Process

The risk assessment process, or in the context of biological agents, the biorisk assessment process, hails from a detailed comprehension of the intrinsic properties of the biological assets. Microbiologists and virologists have long-documented the risks posed by such agents, and their routes of exposure to humans, animals, and the environment. This understanding of the biological agents' properties allows researchers and safety professionals to project the risks associated with various types of manipulations that may occur in a laboratory setting and devise methods to mitigate or minimize these risks. A common example is the use of respiratory protection when handling biological agents that can be transmitted through aerosol exposure. This knowledge extends well beyond the laboratory and highly technical fields of research. In fact, these principles are commonplace even in the public. For example, public spaces often promote pictographic instructions for containing coughs and sneezes during cold and flu season. These techniques are associated with reducing the spread of these viruses, and thereby minimizing exposure to uninfected people.

On the surface, this may not seem like the result of a sophisticated biorisk assessment endeavor. The point remains that the beginning of a robust biorisk assessment process reside in the knowledge of the intrinsic properties of the biological agents themselves.

Figure 1 demonstrates this directionality and the body of knowledge of the biological agents' intrinsic properties that permit risk identification and mitigation. In this example, the risk could be the potential for aerosol generation of a biological agent during an aspiration procedure. The consequence could be the inhalation of the aerosolized agent and resulting infection. The biorisk assessment process would support the identification of this risk, the consequence, the likelihood of risk and resulting consequence, and tactical methods to mitigate that risk.

In short, it is the intrinsic properties of the biological agent itself that drive the directionality of the biorisk assessment process. Clearly this is an oversimplified diagram and process that only serves to point out the foundational elements of a sophisticated process. The biorisk assessment process today is a well-practiced, multifaceted approach to support a wide variety of laboratory operations. It takes into account not only the properties of the biological agent, but also the laboratory procedures, associated laboratory equipment, environmental and engineering controls, likelihood of occurrence, practices, tools, and equipment for mitigating risks, and a determination of risk tolerance [3, 10, 13].

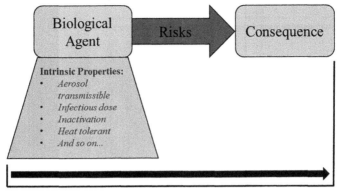

Biorisk Assessment Zone

Fig. 1 Directionality of the biorisk assessment process, defining the biorisk assessment zone. Biosafety combines knowledge of the agents' intrinsic properties, mechanisms of exposure and infection, to identify risks and propose mitigation techniques

3.2   Directionality of the Biothreat Assessment Process

Chapter two "The Biothreat Assessment as a Foundation for Biosecurity" will discuss in detail the creation and use of a threat assessment process specifically adapted for assessing threats and vulnerabilities associated with biological agents (i.e., the biothreat assessment). In the context of biorisk management programs for laboratory operations, like in the biosafety mantra, the focus of the biothreat assessment process is once again at the biological agent level. Only here, the focus has a different perspective. Whereas the risks directed by biological agents form the foundational elements of the biorisk assessment, the biothreat assessment process, and thereby biosecurity, take the perspective of mitigating the threats that are directed *to* the biological agents. Figure 2 demonstrates the overlap of the respective practices of biosafety and biosecurity, yet different perspectives of the same directionality that codifies disparate fields of practice: risk versus threat assessments.

   Another notable distinction is that by analyzing this directional pathway through a lens of threats, it is revealed that biosecurity is focused on an entirely different set of intrinsic properties of the biological agents. The ability of biosafety or biorisk assessment professionals to adequately identify and mitigate risks resulting from the biological agents is a deep comprehension of the *biological* intrinsic properties, such as pathogenicity and duplication rates. Biosecurity programs, established on a foundation of threat assessment and management practices, must also look at the non-biological properties of the biological agents. An asset (biological agent) may not be particularly dangerous insofar as human or animal health impacts, but has tremendous monetary value. For example, recent U.S. Federal Drug Administration-approved biotherapeutics that utilize human gene transfer technology have the ability to correct certain medical conditions with a single dose. These doses, however, could

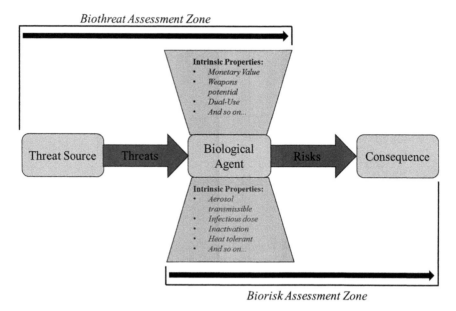

Fig. 2 The biothreat assessment zone is shown as threats being directed toward the biological agents or assets

exceed several hundreds of thousands of U.S. dollars. It is this property, among others, that make this asset desirable and in need of protection from external threat sources.

Of course the dangerous properties of certain biological agents can have an impact from both perspectives of biosafety and biosecurity. It has been well-documented that biological agents have been used as weapons (i.e., bioweapons) and therefore are sought after by individuals or groups with an intention to use those agents, and their dangerous properties, to cause terror or inflict harm [2].

4 Biosecurity is Unique from Traditional Security

Many institutions are challenged to develop and implement biosecurity programs, often because at a funding level, biosecurity programs fall under biosafety programs, or institutions have independent security programs. Often, the argument is be made that institutional (i.e., traditional) security programs should be adequate and sufficient to encompass the needs for security of valuable or dangerous biological materials.

In addition, laboratory programs that are subject to the FSAP, which includes certain fundamentals of securing biological select agents and toxins (BSAT), comply with the laws without extrapolating prescribed biosecurity measures to non-FSAP areas of their programs. Another common scenario is that programs who are not working with BSAT and are not subject to the requirements of the FSAP simply don't develop biosecurity programs at all.

Finally, other programs assume that biosecurity matters can easily cassette into the institutional biosafety program, citing significant similarities and overlaps that would render a dedicated biosecurity program (and/or personnel) redundant.

In all of these scenarios, and others, the assumption has been made that biosecurity is a semi-redundant function of a previously-established program. However, as the directionality discussion above demonstrates, comprehensive programs are built around two foundational elements: biorisk and biothreat assessments. The expertise necessary for the implementation of both does not always reside in a singular office or individual. The intent of biosafety and biosecurity have sufficient disparity, that consideration for building overlapping, yet distinct, programs can be advantageous.

Beginning with the scenario that institutional security is sufficient for implementation of biosecurity, what makes biosecurity unique from a traditional security program? Again, we must look at the intrinsic properties of the biological agents and assets. Further, we must analyze the characteristics and traits of the threats that target vulnerabilities resulting in unwanted consequences to assets.

4.1 Intrinsic Properties of Biological Agents Drive Uniqueness of Biosecurity

In a very oversimplified model, traditional security professionals view the term "threat" as a person who has the intention or ability to inflict harm upon an asset. An asset can be a person (e.g., a political candidate or high-profile CEO) or an object of some particular value (e.g., material for a nuclear weapon, valuable jewels, etc.). This paradigm establishes the protection of the asset from a threatening person. To build a security program, the security professional must understand the properties of the asset to be protected, as well as have as much intelligence on the potential threats [1, 4, 7–9].

The threat assessment process, foundational to a security plan or program, is also the foundation of a robust biosecurity program. A threat assessment is informed by data points on the assets to be protected, the potential threats, the vulnerabilities that provide openings for the threats and assets to interface, and strategies to mitigate those threats and vulnerabilities, or alter properties of the asset. Threat assessment and management are sciences unto themselves, and have been widely employed in a variety of settings, primarily in the context of law enforcement [5, 11]. But the principles of threat and vulnerability identification and mitigation strategies directly apply to the field of biosecurity.

It is worth comparing and contrasting properties between biological materials and other assets. It is this contrast that exemplifies the unique aspects of biosecurity programs versus traditional threat management or institutional security programs. These properties include, among others, the difficulty to detect the asset, the small amounts or volumes of biological agents necessary to cause harm, and the fact that under certain conditions many biological agents can reproduce.

The ability to track or trace an asset is essential to ensuring its security. This is not a difficult concept when the asset is a person, or even a nuclear material. However, unlike nuclear materials, biological materials are not easily detected. While technology is advancing our abilities to detect and signal the presence of bacteria, viruses, and toxins, there is no ubiquitous "Geiger counter for biological materials," or law enforcement K-9 units trained to sniff out these materials. In fact, harmful quantities of biological materials aren't necessarily even visible. This is one of the intrinsic properties of biological materials as assets that make them difficult to protect.

Secondly, tiny volumes of biological assets can still have large impacts, either through causing harm to humans, animals, plants, or the environment, or in terms of monetary value. Therefore, threats with the intent of removing biological materials without detection need only remove miniscule volumes.

Finally, and building on the first two points, many biological materials can reproduce under the correct conditions. This is accomplished through biological replication and propagation. Combined, it is these attributes that separate biosecurity from traditional security programs. An intentional threat (i.e., person) can theoretically remove tiny amounts of dangerous biological material, practically undetected, and reproduce the material to a desired quantity. The challenge for biosecurity programs is to develop and implement strategies and tactics that take into account these properties, built upon traditional threat assessment and management practices.

4.2   The Unique Nature of Threats and Hazards in Biosecurity

The previous section documented unique attributes of the biological agents and assets themselves as potential challenges to building effective biosecurity programs. Based on the model proposed in Fig. 2, the other half of the equation is the threats themselves. As stated, traditional threat management practices identify threats as people, implying intent to cause harm. While this remains true with respect to biosecurity, it is important to expand the threat landscape to include other types of hazards, such as non-malicious threats, weather, information technology, and even negligence.

The possibility of a directed, malicious outsider threat (e.g., terrorist organization, disgruntled employee) is likely the primary threat type that receives the greatest attention, particularly in the fictional media. There are many fictional (and some non-fictional) accounts of such an individual gaining access to a store of dangerous biological material that they intend to use in some malevolent scheme to inflict terror or harm. The reality is that these scenarios are uncommon compared to other types of threats and hazards. Biosecurity programs must account for this possible scenario, however. Therefore, it is important to assess the properties, or attributes, of malicious outsider threats. This book will explore the different types of malicious threats and how those seeking to misuse biological agents do in fact have unique attributes.

More likely than an intentional, directed, malicious threat is the threat that comes in the form of negligence. A simple threat of this nature could be an inattentive laboratorian who inadvertently removes a sample of a biological asset from a secured location to an unsecured location. This is perhaps an individual with warranted access to the assets but with a negligent approach to biosecurity considerations.

Outsider and insider threats are not unique to biosecurity- these are primary threat groups that are the focus of any robust institutional threat management and security program. Arguably, what makes threats in the landscape of biosecurity unique is the knowledge required to use the assets in a manner that is not prescribed by institutional policy, law, or moral context. This book will explore how what used to be a body of knowledge limited to highly-educated professionals is making its way into a broader audience. A prime example of this is the "DIY Bio" movement, or community laboratories [6, 12]. These environments offer training and education, hands-on experience in laboratories, and a lower threshold of access to unqualified persons.

In short, threats can be as unique as the biological assets themselves and require knowledge of their properties and attributes to adequately address and mitigate them.

5 Biosecurity is not Limited to the Laboratory Environment

As discussed above, the inception of the term and implementation of biosecurity, as a body of knowledge and practice, did not begin in the laboratory environment. In fact, restricting the principles of biosecurity to the laboratory alone is likely a limited and damaging view in the context of infectious, potentially dangerous biological agents. It is the intent of this book through the subsequent chapters to reveal how biosecurity can be applied inside and outside of the laboratory. These additional environments will demonstrate a connection and application of biosecurity with respect to public and global health, defense, agriculture, and developing technologies.

The laboratory environment has an obvious need for biosecurity. And there are many types of laboratories, ranging from public health, academic, government, private, industrial, biopharmaceutical, engineering, genomics, and others. Each have their own challenges and a need for an institution-specific biosecurity program. For example, academic laboratories employ a wide range of scientists and students, all with varying access to potentially dangerous or valuable biological materials. The constant flow of staff and students, in an environment designed to be collaborative, can perpetuate gaps in biosecurity program management. Biopharmaceutical laboratories, on the other hand, may not be dealing with agents that are particularly dangerous from a public health perspective, but may have very sensitive operations to mitigate competitive intelligence, intellectual property concerns, and reputational risk.

Outside of laboratories altogether are scores of potentially dangerous biological assets. Disease "hot-spots" around the world occurring through natural circumstances offer unfettered access to dangerous agents. For example, the Ebola virus disease (EVD) outbreak in West Africa during 2014–2015 demonstrated how the need to respond to an epidemic is in contrast to the corresponding need to contain access to infectious materials- in other words, biosecurity and public health can often stand at odds, despite a common good. It is a relatively new concept to consider outbreak environments as assets to be protected from threats, but may support public health and defense efforts in future scenarios.

Finally, we must understand the relationship of biosecurity to biodefense in a comprehensive, applicable manner. Biosecurity is a theory regarding the protection of assets, whereas biodefense is aimed at protecting human, animal, and environmental health from the threat of biological agents. Clearly biosecurity programs inside and outside of the laboratory environment must be considered in the context of a robust biodefense system, but that requires an analysis of specifically what biosecurity means to biodefense and how it can supplement. This book will explore in detail the relationships and the nexus of threat reduction, biodefense, public health, and biosecurity.

6   Biosecurity Extends to Global Health, Defense, and Developing Technologies

The significant point underpinning the content of this volume is that not only does biosecurity begin and end at the laboratory level, but that is has direct implications into other facets of concern, magnitude, and consequence. These include global health, defense, and developing technologies.

Global health generally describes efforts to improve public health at a global level, often but not always, focused on human health issues. However, this is significant appreciation of the interconnectedness across human, animal, and environmental health, often referred to as "One Health." Many factors have a role in the status of public and global health measures, such as prevalence of communicable and noncommunicable disease, economic conditions, basic infrastructure, access to and quality of healthcare services, political conditions, and others. Where biosecurity as a discipline has grown in criticality relative to public and global health issues is squarely in the infectious disease arena.

At the time of this writing we are amidst the greatest pandemic challenge seen in generations: COVID-19. The loss of life, economic collapse, strained health systems, and even changing social and political dynamics, are all evidence of the far-reaching consequences of a planet and society unprepared for the might of an emerging infectious agent. The coronavirus pandemic has laid bare the weaknesses of our health systems, revealed the sensitivity of global economics, and demonstrated that even our social complexities, in terms of how we respond as individuals and populations to disease threats, cannot withstand the impacts of a rapidly-spreading virus. The human race has found itself in a truly defensive position with only lessons to be learned. Perhaps applications of biosecurity can be viewed as one weapon in a developing arsenal to reduce the impacts of the next global pandemic. We will explore this topic in detail (Chap. "Applied Biosecurity in the Face of Epidemics and Pandemics: The COVID-19 Pandemic").

Defense and biodefense are complex topics that require the synergies of global health, policy, intelligence services, threat reduction initiatives, and countermeasures. We find ourselves today not only defending our societies from the threat of naturally-occurring infectious disease but also the threat of intentional acts of terror using biological agents as weapons. As such, biodefense is of growing domestic and international interest and concern, and the relationships of biosecurity and biodefense have not clearly been mapped out. Threat reduction programs, cooperative international support, and the advent of programs such as the Global Health Security Agenda (GHSA) are driving discussions at a global level about how to collectively prepare and respond to both intentional, accidental, and natural biological incidents.

To add to the broader discussion is the advancing march of technological developments. New technologies, such as genome editing, synthetic biology, human gene transfer and therapy, personalized medicine, cyberbiosecurity, and others will force biosafety and biosecurity practitioners to constantly reevaluate the efficacy of their programs.

7 Summary

The roots of biosecurity as a practice can be drawn back to the agricultural sector where protecting livestock from infectious agents was critical. Today, we see the term biosecurity moving into new arenas of practice, including laboratory environments, personalized medicine, agricultural research and production, drug discovery and development, and emerging digital and genomic methodologies.

Further, biosecurity is only recently being critically designed and implemented as a system that is built upon threat management foundations. The lessons learned from law enforcement will continue to shape the design of new biosecurity programs, inside and outside of the laboratory environment. We are only now beginning to recognize biosecurity as a field of practice distinct from biosafety and are considering credentialing programs to recognize biosecurity professionals.

For these reasons and others, we are quite literally redefining the term and practice of biosecurity to meet these growing challenges. It is the intent of this book to display in detail these areas of influence related to biosecurity, and how to apply the practices and principles of threat assessment, management, and biosecurity to a variety of programs and areas.

References

1. Allen G, Derr R (2016) Threat assessment and risk analysis: an applied approach. Butterworth-Heinemann, Amsterdam, Oxford
2. Burnette R (2013) Biosecurity: understanding, assessing, and preventing the threat. John Wiley & Sons, Hoboken, NJ
3. Meechan P, Potts J (2020) Biosafety in microbiological and biomedical laboratories, 6th edn. U.S. Department of Health and Human Services, Public Health Service, Centers for Disease Control and Prevention, National Institutes of Health, Washington, D.C.
4. Deisinger G (2008) The handbook for campus threat assessment & management teams, 1st edn. Applied Risk Management, Stoneham, MA
5. Fein RA, Vossekuil B (2000) Protective intelligence and threat assessment investigations : a guide for state and local law enforcement officials. U.S. Department of Justice, Office of Justice Programs, National Institute of Justice, Washington, DC
6. Keulartz J, van den Belt H (2016) DIY-Bio - economic, epistemological and ethical implications and ambivalences. Life Sci Soc Policy 12(1):7. https://doi.org/10.1186/s40504-016-0039-1
7. Meloy JR, Hoffmann J (2014) International handbook of threat assessment. Oxford University Press, Oxford
8. Meloy JR, Sheridan L, Hoffmann J (2008) Stalking, threatening, and attacking public figures : a psychological and behavioral analysis. Oxford University Press, Oxford, New York
9. Merkidze AW (2007) Terrorism issues : threat assessment, consequences and prevention. Nova Science Publishers, New York
10. Salerno RM, Gaudioso J (2015) Laboratory biorisk management: biosafety and biosecurity. CRC Press, Hoboken
11. Scott-Snyder S (2017) Introduction to forensic psychology : essentials for law enforcement. CRC Press, Taylor & Francis Group, New York
12. Seyfried G, Pei L, Schmidt M (2014) European do-it-yourself (DIY) biology: beyond the hope, hype and horror. BioEssays 36(6):548–551. https://doi.org/10.1002/bies.201300149
13. World Health Organization Department of Epidemic and Pandemic Alert and Response (2004) Laboratory biosafety manual, 3rd edn. World Health Organization, Geneva

The Biothreat Assessment as a Foundation for Biosecurity

Ryan N. Burnette and Chuck Tobin

Abstract Threat management as a system or theory, and practice, is rooted in the ability to systematically identify potential threats and mitigate the risks of these threats acting upon a particular asset. Threat assessment is not a snapshot in time but a living process to detect threats as early as possible to facilitate mitigation and management of them. As this chapter will discuss, the foundations of threat assessment and management as a body of expertise draws its history from law enforcement, security, and the behavioral sciences. The outputs of this history include a variety of tools, from behavioral analysis to discrete security protocols that continue to demonstrate their effectiveness in the related fields of investigations, security, defense, and personnel protection. Despite the significant differences between the assets in a traditional threat management perspective (e.g., human assets/targets) and the biosecurity perspective (e.g., biological agents, toxins, animals, technology, data) tenets of threat management can be effectively applied to the fields of biosecurity, if not underpin it. This chapter will explore the modification and adaptation of the threat management backbone tool, threat assessment, into a biosecurity specific tool. Effectively, this chapter will describe the resulting *biothreat assessment* process.

1 Threat Management as a Field of Practice in Security Programs

Threat management is a burgeoning field unto itself, more so outside of the life sciences than in. The codification and dissemination of threat assessment and management as a distinct field of practice draws primarily from law enforcement but is currently finding broader applications. With respect to the life sciences, we

R. N. Burnette (✉)
Merrick & Company, Washington, D.C., USA
e-mail: ryan.burnette@merrick.com

C. Tobin
AT-RISK International, LLC, Chantilly, VA, USA

© Springer Nature Switzerland AG 2021
R. N. Burnette (ed.), *Applied Biosecurity: Global Health, Biodefense, and Developing Technologies*, Advanced Sciences and Technologies for Security Applications, https://doi.org/10.1007/978-3-030-69464-7_2

13

tend to think of the risk assessment process as a systematic qualitative or quantitative approach that allows professionals to identify, mitigate, and manage consequences that are presented by biological materials, such as an infectious agent. Specifically, a biological risk, or biorisk, is the probability of an adverse event stemming from a biological agent. The practices and principles of managing these biorisks is simply *biorisk management*. But as this chapter details, biorisk management is also dependent upon the "other half of the equation," or *threat management*. Threat management specifically deals with the threats and hazards directed at the biological agent or asset of value or consequence. To understand the role of the threat assessment process in the context of the life sciences, it is important to understand the fields of threat assessment and management.

1.1 Defining Threat Assessment

Interpretations of the term "threat assessment" tend to be broad and wide. Ranging from the definitions promoted by ASIS International in their Risk Assessment standard stating that threat assessment is "the process of identifying and quantifying the potential cause of an unwanted event which may result in harm to individuals, assets, a system or organization, the environment or community" [1]. To the more specific version published by the U.S. Federal Bureau of Investigation (FBI) in "Making Prevention a Reality" and similarly followed by the Association of Threat Assessment Professionals" (ATAP) which is more directed towards targeted threats against an individual or organization [8, 2]. Their definition, "threat assessment is a systematic, fact-based method of investigation and examination that blends the collection and analysis of multiple sources of information with published research and practitioner experience, focusing on an individual's patterns of thinking and behavior to determine whether, and to what extent, a person of concern is moving toward an attack." In the conventional sense, threat assessment is part of a process within risk assessment to help us identify the likelihood of threats occurring.

Threat assessment is not and should not be a singular point in time. A onetime programmatic review that somehow informs stakeholders of a static condition is not useful. Threat assessment is constant and dynamic, mostly driven by human actors who may intentionally or unintentionally enhance the risk to an organization or individual. Applying threat assessment within the context of the biosecurity environment presents some unique considerations. Not only is the assessor considerate of the consequences of a particular event such as sabotage, vandalism, or theft of an asset, but they must also recognize that the impact of theft of controlled materials or agents may have a much greater impact on the community at large. The assessor must be aware that individuals seeking to compromise a biological material may not be doing so for personal gain. For example, their motivation may be to affect mass losses of life as a result of obtaining unauthorized access to the materials and weaponizing them.

1.2   History of Threat Assessment

In order to properly frame threat assessment and facilitate the evolution of tools for biosecurity programs, we should first review the history of threat assessment that has taken us to today. The model that has evolved has actually come from two distinct tracks of professional communities. Threat assessment in the context of identifying and reducing the likelihood of incidents of crimes against persons and property grew out of law enforcement and security drafting best practices to mitigate threats. These models evolved thanks to input from professional associations such as the International Association of Chiefs of Police and ASIS International. Dialogue surfaced suggesting designing concentric circles around an asset, effectively layering protection to delay and detect a threat coming in proximity to the asset. The production of theories such as Crime Prevention Through Environmental Design (CPTED) addressed threats through community design. Over time, the model ASIS International Risk Assessment standard was produced with heavy influence from the ISO-3000 standards. Originally these models were heavily focused on crime prevention through a presence that would deter interest in an asset by placing guards, systems, and barriers in place.

Similarly, theories of threat assessment in the behavioral context date back decades but found their foundations in traditional violence risk assessment, primarily focused on incarcerated individuals pending release back into society. These static assessments performed by clinicians informed the courts on whether an individual would present a significant risk of violence upon their return to society. This practice evolved over time and even law enforcement's view on violence prevention has become less reactive and more focused on addressing behaviors before they become criminal acts requiring investigation. It is well described by Borum, Fein, Vossekuil and Berglung in *Threat Assessment: Defining an Approach for Evaluating Risk of Targeted Violence*: "Conceptually, there has been a shift from the violence prediction model, where dangerousness was viewed as dispositional [residing within the individual], static [not subject to change] and dichotomous [either present or not present] to the current risk assessment model where dangerousness or "risk" as a construct is now predominantly viewed as contextual [highly dependent on situations and circumstances], dynamic [subject to change] and continuous [varying along a continuum of probability]" [3]. This team went on to further produce the Exceptional Case Study Project as well as the Protective Intelligence and Threat Assessment Investigations guide which helped further frame the assessment of threats. While their design was oriented towards helping law enforcement form protective intelligence and threat assessment programs to protect public figures, the take-aways overlap our needs.

Taking both the history of the law enforcement/security and clinical models of threat assessment we can see where the opportunity exists to merge both approaches into a model that is proactive and comprehensive. It goes beyond static assessments and becomes a dynamic model that adjusts to address threats facing the organization.

1.3 Framing Threat Assessment

Older approaches to threat management harken of medieval castles with moats and sentries, raising bridges and lowering gates to protect the community from onslaught by the savages to newer models where the king would dispatch intel sources into the woods to identify where the threat was headed and if possible, influence a change in thinking. Similarly, modern threat assessment programs shouldn't simply exist to wait for the threat to present itself. Programs must constantly look for the threat and adapt to the ever-changing threat landscape. We have seen this best displayed by the evolving war on terrorism. Each time we design a mitigation strategy to the threat presented, they identify the next vulnerability and, repeatedly, we find ourselves on our heels.

Taking a broad look at threat assessment, we seek to identify threats arising from human actions/inactions, technological events, natural threats, and other occurrences (Fig. 1). This book will not expand on natural threats or technological threats except

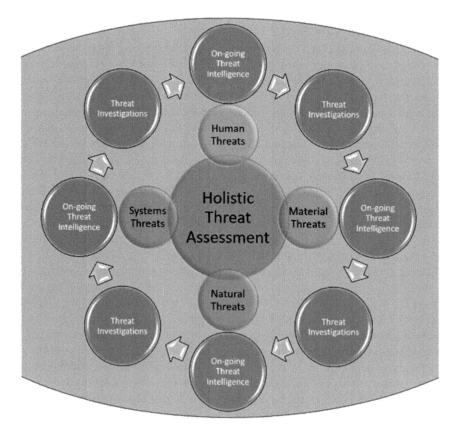

Fig. 1 A holistic threat assessment model is comprised of many tools and processes

to note that mother nature and systems failures should not be omitted in the evaluation of threats that may impact your risk. Focusing further on human actors we can identify a myriad of considerations. Threats from insiders and outsiders, random or targeted acts as well as events associated with non-compliance or negligence must be considered. As a result, a threat assessment program must be created to identify potential threats facing the organization but also continue to evolve and inform a process that is actively detecting and evaluating new threats. With the objective of understanding the capabilities and motivation behind attackers, we should recognize the level of sophistication necessary to deploy an effective threat management program. Many would argue that understanding motivation is irrelevant to designing security mitigation strategies, because just knowing their capabilities doesn't give us enough information to be able to influence change or redirection. This is especially important when we are looking at targeted actions by an insider that may commit acts of workplace violence, theft, or sabotage.

A traditional threat assessment performed by security practitioners typically results in some qualitative or semi-quantitative value such as high, medium, or low threats and many times is then used to inform a larger threat assessment [1]. A sample of such a threat assessment may evolve as follows:

"Consultant evaluated local uniformed crime report data and has determined that threats associated to arson at your facility is LOW due to…."

This model approach is commonly employed to help inform decision makers during the risk assessment process to prepare the budget for enhancements to a developing or existing program. A comprehensive threat assessment program needs to go beyond this static model and continually collect intelligence to inform the threat assessment program. The threat assessment program becomes a living program, regularly informed of the data and statistics that impact the overall risk, but also collects

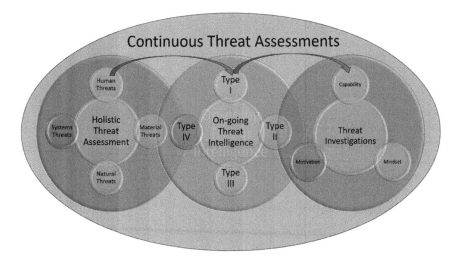

Fig. 2 Threat assessments are conducted in a continual model

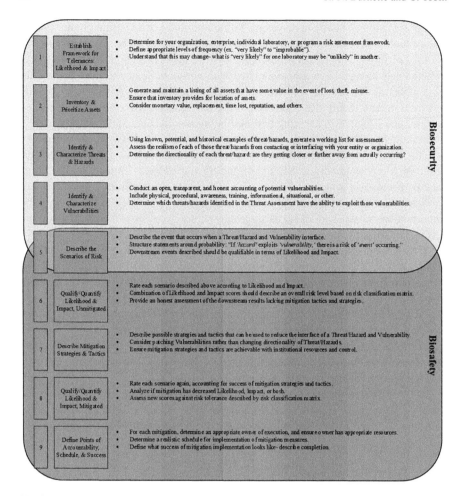

Fig. 3 Nine-point process for conducting the Biothreat and Biorisk Assessment

intelligence relative to specific threats that impact the organization. A dynamic threat assessment model promotes evaluating individual and group motivations and capabilities to perform an unwanted act against the organization while actively investigating these persons or groups to identify movement from an idea to action. As noted earlier, these actors may come from inside or outside of an organization. The National Institute of Occupational Health and Safety even provides a map to work within when assessing threats for violence against people. They classify threats of violence in the workplace via four categories.

- Type I—violent acts by criminals with no other connection to the company
- Type II—violence directed at employees by customers, clients, patients, students, etc.

- Type III—violence committed against employees, supervisors, or managers by current or past employees
- Type IV—violence committed in the workplace by someone who doesn't work there but has a personal relationship with an employee

It is reasonable to consider these as a template for targeted threats from persons with no employment connection, persons with a customer/vendor relationship, persons with an employment related relationship and lastly those with an intimate relationship with persons employed by the target organization.

ASIS International, in its Risk Assessment guide suggests three possible tree models for threat assessment. The asset tree where an analysis of the asset, means of access, internal or external threat actors, intentional or unintentional motive, capability, event and consequences results in a threat summary. A second tree they propose suggests a threat type tree where the analysis looks at the type of threat, act, resultant event and consequence. Lastly, they outlined an adversary tree analyzing the type of adversary, motivation, capability, methods, events and consequences [1].

1.4 Threat Assessment Approaches and Models

As outlined above, many tools and models are available to assist with a general threat assessment conducted in conjunction with an organization's risk assessment [1]. This section will consider models designed to identify behaviors of concern and risk factors that elevate an individual or group's propensity for violence. In reality, using these tools, we may identify not just an interest in violence but interest in committing a multitude of unwanted activities such as sabotage, vandalism, theft of trade secrets, and others. By considering individual or group activities of concern and behaviors of concern, we can create a form of early warning system that may suggest that the threatening activities are emerging. These behaviors and activities are visible to the public and over time accumulate to inform us of an individual's movement along a pathway towards violence or other unwanted activities.

The evolution of fears resulting from radicalized individuals committing targeted acts of violence has also driven the creation of investigative tools such as the TRAP-18 [9, 12]. This tool identifies eight (8) proximal warning behaviors and ten (10) distal characteristics of a person of concern. Consider proximal warning behaviors as those indicators that demonstrate an individual is now moving along a selected pathway towards violence, whereas the distal characteristics provide more of an outline of risk factors for radicalization. Scoring these various factors such as an individual's demonstration of a personal grievance, framed in an ideology with a demonstrated dependence on a virtual community and recent thwarting of occupational goals (all distal characteristics) coupled with an abnormal fixation and identification with assassins or the "evil hero" may suggest an individual presents a greater risk.

The question becomes, how do we take models that are oriented towards a broader threat scape and align them with tools that function within a living and breathing threat scape of targeted acts against an organization? Utilizing the model earlier as reference, each threat type in a holistic threat assessment model is supported by ongoing intelligence and investigations of threats. Within the category of human threats, we would evaluate Type I, Type II, Type III and Type IV matters seeking to identify capabilities, mindset and motivation. This model, adopted across all aspects of threat assessment would allow organizations to recognize the threats facing the organization in perpetuity and facilitate the evolution of threat management. Matters to include personnel screening, insider threats, interpersonal violence such as domestic violence and many other threats would be ferreted out through the evolution of the program.

2 Biosecurity Threats and Hazards

The biosecurity environment is subject to a myriad of threats. Some are unique to the environment, where others are common in many industries, yet often overlooked. No different than the typical corporation, hospitals, research centers, pharmaceuticals, or others all may suffer from the likelihood of incidents to include theft of assets, sabotage, workplace violence and vandalism. Many institutions have taken novel approaches to identify and address threats facing them, however, the vast majority have fundamentally inadequate models. These organizations need to review the threats facing them and examine their own unique vulnerabilities and mitigation strategies within administrative, operational and technical controls. Chapter one "Redefining Biosecurity by Application in Global Health, Biodefense, and Developing Technologies" provided insight into the uniqueness of biosecurity programs based on the intrinsic properties of the assets being protected, guarded, or monitored. Another aspect that makes biosecurity a unique subset of the security industry is the nature and types of threats and hazards.

2.1 Bioeconomy Today Presents Novel Threats and Hazards

Examining the bio-community we encounter new threats that are not likely addressed by typical security programs. Considering that laboratories, hospitals, storage facilities, production environments, and biorepository facilities alike all possess materials that individually may present a threat in and of themselves. The threat assessment needs to consider not only the consequences of loss of an asset, such as a valuable biological material, but the impact that asset may have on environmental or community security matters. Having possession of an item that is desirable as a potential means of advancing science by others or facilitating mass illness and/or death presents a unique challenge. Not only should your program consider the theft for an item's "cash" value, but you also are faced with the item representing competitive advantage and possible use as a weapon. Perhaps compare certain items as similar to nuclear weapons, others as works of art, and others as critical to the continuity of a project.

Taking our approaches outlined earlier, we can look at our threats from the perspective of the threat type, asset type, and adversary type. Considering an adversary type of threat assessment and many ways an individual may achieve a single objective of obtaining access to biological agent:

- Obtains a direct hire position within the laboratory
- Obtains a vendor/contractor hire position with proximity to the laboratory
- Obtains an internship through a sanctioned educational program
- Participates in sanctioned tour of facility under ruse
- Bypasses existing security controls through bribery, coercion or blackmail
- Compromises the facility through an orchestrated attack
- Compromises material in transit through an orchestrated attack

This limited list explores the many pathways an individual could gain access to sensitive assets. Many of these pathways would remain covert and likely undiscovered without a sophisticated screening program that actively engages a threat intelligence program, usually operated by law enforcement, federal, or major corporate entities.

We would be remiss in addressing threats and vulnerabilities if we didn't consider threats associated with negligence or incompetence. An individual's inability to follow proper protocol doesn't diminish the threat presented by certain material components. They still can be compromised through a negligent action and cause significant damage to the organization, its brand or even the community. Consider the technicians lab coat and the best place to warm a tube. Distracted and perhaps overwhelmed by their workload and demands to produce research by deadlines that are fast approaching results in forgetting to place a tube back in its original condition. The coat departs the facility without inspection and ends up in the laundry at home. Your toddler nephew visiting your home discovers the lab coat and while playing doctor discovers the vial of superhero formula and consumes its contents.

Perhaps this science fiction story is a little out over the top, or is it? Perhaps we can look at similar outlandish stories to engage in fantasy for a moment. Your new intern has shown a limited capacity to perform, but he is trying. His grades suffer, partly due to incompetence but largely due to his fixation of another female intern. As his fixation becomes obsessive you recognize a deterioration in his performance and focus. Simple mistakes are becoming costly in time and materials. You take affirmative action to discipline this individual in the lab. Feeling shamed and now further distanced from his future career and more importantly since his true love witnessed this disciplinary activity, she appears to increase her distance from him. When in reality, she was never connected to him but engaged to be married to another man. Distraught over this evolving situation his grievance of a lack of respect and loss of his love results in the evolution of an ideation: "If I can't have her, no one will." So, one night while working in the lab alone he manages to gain access to his love without disruption. In his last-ditch effort to win her love he is dismissed completely by her. In retaliation and response to his humiliation he turns to rage and murders her in the lab, burying her body inside the walls of the lab. While not a precise documentary of the murder, the outlined story is not too far from a real-life situation that occurred [10].

2.2 Gap in Expertise Lowers Threshold to New Types of Would-Be Actors

The selection of a weapon as a means to manifest fear and terror, seek monetary gain, or seek other goals, is often tied to the desired outcome [11]. As a simple example, hackers seeking to hold ransom a company's data will use ransomware, computers, and networks as their "weapons." Threat assessment processes also tell us that malicious actors have reasons for being drawn to the types of weapons they choose. These reasons may include level of access or physically being able to gain access, intended outcome (e.g., violence, fear, murder, mass casualty), and expertise to manipulate the weapons to be effective toward the goal. In fact, comparisons of various attack methodologies, intents, and weapons, reveal that biological agents have been used in a staggeringly small percentage [11]. Data and studies such as these suggest that the criteria used for weapon selection are less attainable. Is it possible that biological agents as weapons (bioweapons) have not been observed as frequently as firearms, explosives, or chemical agents because of the perceived level of expertise needed to effectively wield them? Data are lacking to confirm or deny that unique types of malicious actors are drawn specifically to bioweapons [7].

Despite large, organized biological weapons programs in history, the use of bioweapons for small acts of terror are limited. Perhaps one reason for this observation is the perceived level of expertise necessary to effectively manipulate, cultivate, and deploy biological agents is too high for most malicious actors. But as technology continues to advance, it is reasonable to assume that the expertise threshold will continue to decrease. Subsequent chapters will go into detail on the technology aspects that are opening doors to many would-be biologists, or "biohackers," and that this bears some concern over increasing access to biological tools that could be used as weapons.

Firearms, explosives, and nuclear devices have rapid and obvious impacts when used as weapons. The use of biological agents is perhaps more subtle and drawn out over a period of time. Another possible reason that bioweapons account for so few attacks is whether they are effective at eliciting the desired effect of fear, terror, or death. Perhaps malicious actors and policy makers both have underestimated the consequences of biological agents as weapons [4]. And perhaps the coronavirus pandemic has changed that perception. The widespread, global, and significant damage the COVID-19 pandemic has rendered is staggering. It is arguable that this naturally occurring virus has delivered stark evidence that biological agents have the ability to level economies, create global fear, and result in tremendous loss of life.

Will this pandemic lower the bar for individuals or organized groups to attempt the use of biological weapons? In combination, access to biological agents, and developing technologies to manipulate them, and evidence of the impacts of a novel virus, all coalesce to depict a changing threat landscape.

3 The Biothreat Assessment as a Novel Approach to Combine Science and Threats

A major intent of this book is to describe a novel threat assessment tool and process specific to the field of biosecurity, or the *biothreat assessment*. A primary challenge in this process is to capitalize on the foundations prescribed by the law enforcement and clinicians, but to step away from the boundaries of the traditional definition of a "threat" being limited to a person or people that we react to. Therefore, this biothreat assessment process has been expanded to include additional hazards that can threaten the integrity or security of biological agents and assets and to assist in formulating proactive mitigation strategies.

Additionally, the biothreat assessment must provide a logical, directional link back to a system of biosafety considerations. Review of Fig. 2 (chapter one "Redefining Biosecurity by Application in Global Health, Biodefense, and Developing Technologies") reveals that (1) a threat or hazard can accomplish access to a biological asset and (2) this compromised access can result in an undesirable, resulting risk. In this simplified anecdote, a malicious outsider threat (e.g., a former laboratory employee) uses his or her credentials to illicitly access a dangerous biological asset (e.g., a harmful bacterial strain), removes it from the laboratory and uses it to inflict harm (e.g., exposes the agent to a person or animal). It is this overlap that supports the cross-pollination of biosafety and biosecurity resulting in a more comprehensive biorisk management system. So, the biothreat assessment provides a tangible link back to biosafety programs.

Further, as described briefly in chapter one "Redefining Biosecurity by Application in Global Health, Biodefense, and Developing Technologies" and more detailed above, a biothreat assessment must take into account the unique, intrinsic properties of the biological materials and assets themselves. This is necessary because the resulting risk has the ability to be biological agent specific.

A major hinderance of identifying, mitigating, and managing threats to life sciences has been the lack of an assessment platform that brings together the risk-based (i.e., biosafety) and threat-based (i.e., biosecurity) tools. The ability to draw from the entire continuum of threat to agent, to risk, to consequence (see Fig. 2, chapter one "Redefining Biosecurity by Application in Global Health, Biodefense, and Developing Technologies") is what results in comprehensive biorisk management. Therefore, the fundamental tools, such as the risk assessment and threat assessment, must also be linked to ensure continuity. It is the linkage between these two fundamental assessment programs that biorisk assessment and management is achieved.

3.1 Linking the Risk and Threat Assessments for Biorisk Management Program Continuity

The risk assessment process in the life sciences is a well-documented process that has been used widely to support research, production, medicine, and others [6, 13, 16]. Core to this process is a fundamental and comprehensive understanding of the properties of the biological agents(s) (see chapter one "Redefining Biosecurity by Application in Global Health, Biodefense, and Developing Technologies"). But these processes generally fail to address the threats that may be directed to the programs or agents of concern. That is accomplished through the threat assessment.

The first challenge is to distill a threat assessment process that is largely derived from law enforcement and adjust it into the context of life sciences programs. For example, the asset in question (i.e., the biological materials being protected) is biological in nature. The true threat assessment perspective usually assumes the asset is another human.

Secondly, the adjusted threat assessment process must be linked with the biorisk assessment process. It is this linkage that provides the user or implementer of the assessment process to observe the continuity and spectrum presented by Fig. 2 in chapter one "Redefining Biosecurity by Application in Global Health, Biodefense, and Developing Technologies." The criticality of linking the assessment processes ensures that neither is providing information regarding identification, mitigation, and management are done without continuity. In other words, a threat that impacts the integrity of a biological asset could in fact promote downstream negative consequences resulting from the intrinsic properties of the asset itself.

3.2 Design and Execution of a Linked Biorisk and Biothreat Assessment

To achieve a dynamic viewpoint of the threats, hazards, vulnerabilities, and risks associated with life sciences programs it is important to establish a clear linkage between the commonly practiced biorisk assessment and the developing biothreat assessment processes. This section will provide a simplistic yet effective method to link the threat assessment and biorisk assessment processes through a nine-point process as defined in Fig. 3.

Frequency	Impact			
	Catastrophic (4)	Critical (3)	Marginal (2)	Negligible (1)
Frequent (5)	A	A	A	B
Probable (4)	A	A	B	C
Occasional (3)	A	B	C	C
Remote (2)	B	C	C	D
Improbable (1)	C	C	D	D

Class	Tolerance
A	Intolerable – must be corrected
B	Undesirable – only acceptable when risk mitigation is impractical
C	Tolerable – with endorsement of institutional approval
D	Tolerable – with endorsement of lower levels of approval

Fig. 4 Hypothetical risk classification matrix with tolerance descriptions

Establish Framework for Tolerances: Likelihood and Impact.

The risk classification matrix is a ubiquitous element that underscores many risk assessment methodologies. It provides users and institutions the ability to think about downstream consequences in terms of "likelihood" of occurrence and the "impact" such a consequence will have. An example risk classification matrix is presented below in Fig. 4.

The labels of frequency, impact, and tolerances are all institution specific, requiring reflection of institutional stakeholder inputs. The scale should be adjusted to provide scalability for inclusion of any manner of threats, hazards, and risks.

Inventory and Prioritize Assets:

Organizations should be able to prioritize the value of the assets. In the life sciences, assets can comprise almost anything of value, including biological materials, intellectual property, laboratory equipment, animals, data, historical knowledge and expertise, and of course the human resources. There are many criteria by which an entity may choose to categorize and prioritize assets. Normally, these include monetary value, difficulty to replace, time lost, and reputational damage. However, specific to biological assets that have the ability to pose a health risk to people, animals, or the environment must also be considered. For example, the monetary value of a bacterial strain may be insignificant and yet have the ability to cause significant health damage to people in an uncontrolled environment outside of the laboratory. Therefore, a key component in prioritizing biological assets, through a perspective of biosecurity, is the ability to cause harm. This represents a key cross-over point between biosecurity (protecting the biological asset…) and biosafety (… from harming people, animals, or the environment). Each entity should devise the criteria for prioritizing assets and maintain an active inventory of all assets.

Identify and Characterize Threats and Hazards:

The threat assessment process begins with the identification of a potential or known threat or hazard. As the tools above have referenced, a significant component is characterizing and qualifying the threat or hazard. For most purposes when designing a biosecurity program, it is acceptable to use hypothetical threats and hazards, such as a malicious outsider/insider, negligent insider, non-human hazard (e.g., hurricane, earthquake), or other threat/hazard element. Further, there are a number of law enforcement resources (e.g., Bureau of Justice Statistics; U.S. Federal Bureau of Investigation; Interpol) that provide crime statistics in a given area- these too can be considered as a starting point and are worthy of monitoring periodically. As a hypothetical example we will focus on a malicious outsider. This could be a member of an activist organization, the ex-boyfriend of a laboratory employee, or some other motivated individual. Here, we will consider the individual to be the ex-boyfriend of a laboratory employee.

Because assessments are meant to be dynamic, the directionality of the threat/hazard can change. Recall that threats/hazards moving towards (e.g., the malicious ex-boyfriend has been spotted in proximity to the laboratory building) or away (e.g., the malicious ex-boyfriend relocates across country) has a bearing on subsequent steps of the analysis with respect to likelihood. Imploring behavioral based threat assessment practices suggests we not just review the individual motivations but also their behaviors, intentions, and capabilities. Capabilities may be well illustrated through the earlier outline of the ex-boyfriend. While they may possess the motivation and intentions to cause harm, their proximity, or lack of, dilutes their threat level. Unless, of course, a history or capability of travel are present. A thorough threat assessment will work to provide an understanding of the factors that qualify a threat or hazard and to determine the directionality. Understanding the origin of the threat or hazard can have some bearing on the choice of downstream mitigation tactics that are chosen.

Identify and Characterize Vulnerabilities:

Threats and hazards often have the capability of exploiting more than one, or many, vulnerabilities. Therefore, a threat assessment is most useful when coupled with a thorough vulnerability analysis. A vulnerability analysis is an undertaking to inventory all potential weaknesses in a system, from physical to procedural, that can be exploited by a threat or hazard. Appropriately assessing the vulnerability suggests that the assessor also needs to understand how attractive the particular target may be and the potential consequences of loss. Following research presented by the American Petroleum Institute in their 2003 publication "Security Vulnerability Assessment Methodology for the Petroleum and Petrochemical Industries," we can consider the likelihood of an event occurring is a function of the attractiveness of an asset, degree of threat, and degree of vulnerability. The more attractive the asset and the greater the impact it offers, the greater the likelihood vulnerabilities may be exploited. The higher the skilled adversary may suggest fewer vulnerabilities are necessary for a successful campaign.

In the example of the disgruntled ex-boyfriend, he may choose to exploit the fact the laboratory only has manned security forces present during normal business hours, or the fact that the security camera over the rear entrance appears to be broken, or perhaps he is friends with a separate individual that does have permitted access to the facility. Each of these represent a potential vulnerability that a malicious actor may choose to exploit singularly or in combination. Other examples of common vulnerabilities include out-of-date anti-virus software on networks, lack of defined policies, procedures, and training programs, aging critical engineering controls (e.g., backup power generator), and lax access controls. True to most risk assessment frameworks, vulnerabilities can usually be assigned in one or more of the five major pillars of security: physical, material, informational, personnel, and transport [5]. Understanding how vulnerabilities fit into these categories can be especially useful in later steps that describe mitigation measures and accountabilities.

Describe the Risk Scenarios:

Scenarios allow threat and risk assessment practitioners the ability to forecast potential events, prioritize them based on likelihood of occurrence and impact, and develop mitigation strategies and tactics to reduce either or both likelihood and impact. Scenarios allow users to tell a story with hypothetical threats (negligent insider, for example), hazards (impending hurricane), and vulnerabilities (lack of emergency protocol training). In a way, scenarios are built by playing "match-maker" with threats/hazards and vulnerabilities. Assuming the previous steps in identifying and characterizing threats/hazards, and vulnerabilities have produced a robust list, plotting scenarios stokes the creative aspects of the process. For example, a hurricane may be headed toward an animal laboratory. A negligent employee may be aware of, and feel rushed by, the incoming storm but due to lack of emergency preparedness training, he/she may not properly account for the well-being and security of a population of laboratory animals. The potential consequences could result in loss of the animal colony, unintentional release of animals, or even permittance of outsiders to gain entry. It is up to the stakeholders administering the assessments to determine the range of reality, or probability, of a scenario for being considered. Because each threat/hazard can elicit numerous vulnerabilities, and each vulnerability can produce numerous downstream risks, the list of permutations can grow quite long and prevent the assessment process from being practical.

It is useful to describe scenarios in the form of "If" and "Then" statements. For example, "If an external security camera is visibly broken, then there may be a higher risk of a malicious outsider choosing to break into the building at that location." In reality, most scenarios are much more complex but the ability to form simple statements can help assign each component of the process. For example, in the statement, "broken security camera" represents a vulnerability, "malicious outsider" represents the threat, and "the resulting break-in" is the adverse event. Therefore, these simple "if" and "then" statements provide a simple framework for stakeholders to properly categorize threats, hazards, vulnerabilities, and risks.

Qualify or Quantify the Unmitigated Likelihood and Impact of Each Scenario:

Each scenario should conclude with the stakeholders coming to a consensus on the likelihood and the impact of the scenario occurring. In a way, this is asking "how real" is the combination of events? This is the process where the risk classification matrix becomes useable (see Fig. 4). Classification of each resulting risk should be identifiable in terms of one of the categories of frequency of occurrence and potential impact. The combination of these will reveal the associated risk class. It is important for this step to be done assuming no mitigation strategies or tactics have yet been employed. Therefore, the resulting risk classification represents an accurate appraisal of real, yet potential, events. It is often useful to use a historical example to test the efficacy of the risk classification matrix. It is also common for risk classification matrices to be amended. The point is that these processes are adaptable- they must be able to be adjusted as new scenarios are presented, changes to operations occur, or actual events unfold that provide for feedback learning.

Describe Mitigation Strategies and Tactics:

Up to this point in the nine-point process the focus has been on describing the threats, hazards, vulnerabilities, and their intersections which result in risk. Further, risk classifications have been assigned to each of the scenarios against site-specific definitions of tolerance. It is likely that many scenarios have resulted in the potential for adverse effects that are above the tolerance levels determined. Like the creative step of designing scenarios, this step also involves creativity and knowledge of institutional resources. At this step, stakeholders should design practical risk mitigation strategies and tactics working within the reasonable boundaries of institutional resources. Simply, this means that mitigation strategies and tactics that require unrealistic amounts of time, money, or other resources out of reach to the institution are not considered practical.

Qualifying the unmitigated consequences has hopefully also revealed which scenarios are a priority to mitigate based on likelihood and impact. These priorities will now be tested against the practicality of the mitigation measures. Another important consideration is where the institution or stakeholders have the ability to affect change. Most institutions do not have the ability to alter the nature or directionality of a threat or hazard. Changing the course of a hurricane is not possible. Subverting the ideology of an animal rights activist organization is not possible. The result is that asserting control over the points of vulnerability often represent the best areas to affect change and implement mitigation measures. This can include replacing the broken security camera, reevaluating the need for 24/7 manned security, increasing the frequency of training on laboratory security policies, or adding locks to vessels of sensitive biological assets. The result of this step is to have a defined, practical mitigation strategy or tactic against each scenario.

Qualify or Quantify the Mitigated Likelihood and Impact of Each Scenario:

Prior, we had qualified or quantified the likelihood and impact of each scenario in the absence of mitigation measures. We now will revisit those same scenarios under the assumption that mitigation measures have been implemented. Stakeholders should now objectively evaluate the introduction of mitigation measures to alter the directionality of a threat or hazard, augment an identified vulnerability that reduces the ability of a threat or hazard to interface, and ultimately reduce the likelihood or an adverse event occurring, the impact of negative consequences, or both. Revisiting the risk classification matrix should reveal a decrease in magnitude of either frequency, impact, or both. Resulting risk tolerances can be observed and adjusted. In the end, stakeholders will need to determine if they will in fact mitigate the risk through administrative, technical, or physical solutions, transfer the risk to an insurance program, tolerate the risk, or ignore the risk all together. The last two options likely come with difficult decisions relative to public safety risk, brand risk, as well as possible criminal and civil liability.

Define Points of Accountability, Schedule, and Success:

The assessment process is not complete until mitigation measures have been slated for action. Each discrete mitigation measure should be documented in the form of a task or action, and assigned to a point of accountability, or "owner." The owner has accountability to see the action through to completion and maintains a clear understanding of what success represents. Further, defining a schedule for implementation of the mitigation measure is necessary for actionable outcomes. For example, replacing a broken security camera may be the responsibility of plant maintenance, success is defined as having a functional security camera in its place, and a schedule for replacing the security camera is defined.

Assessments are not useful if they simply represent a singular point in time. Rather, a hallmark of a robust threat and risk assessment process is understanding the milestones to revisit, revise, and adjust. The threat assessment process is a living aspect of the biosecurity program. While each institution should understand what catalyzes revisiting the institutional threat assessment process, be it a change in research programs, updated crime statistics from local law enforcement, an actual incident or "near miss" that provides real-world data, or some periodic timeline for review. Stakeholders must recognize that ongoing threat assessments may occur regarding individuals or groups of concern and their potential impact on the threat landscape. All too often organizations interpret the risk assessment process as a one-time event. Perhaps the larger institutional assessment of threats, vulnerability, and risk may only be performed every year or based on new criteria. The process of conducting behavioral based threat assessments may occur constantly and triggers to inform the supporting mechanisms are essential to identify early warning signs of insider threats, targeted external violence interests, and the like.

3.3   Assessments Represent One Component of a Robust Biosecurity Program

It is important to note that the biothreat and biorisk assessment methodology described above is but one component of a robust, integrated biorisk management program. It does not solely comprise the program. As with any management system, many components with an array of subject matter stakeholders are required. Biosecurity programs are most successful when they include a mechanism for program management against an established set of policies and backed by authorities, physical security measures such as access control, a management process for personnel and their suitability to have access to valuable or sensitive materials, a robust mechanism for maintaining accurate inventories of valuable or sensitive materials, proper frameworks for securing information relative to institutional assets and programs, and adequate policies and procedures that specify how valuable or sensitive biological materials are transported. Further, as with any effective management system, stakeholders should design biorisk management programs with a method of measurement to metric activities and success. Biosecurity program components are well-described and widely available for referencing [5, 6, 13]. These biosecurity program elements, using the biorisk and biothreat assessment methodology described in this chapter, represent the ideal framework for institutions to adopt, adapt, and implement.

4   Connecting Threats/Hazards, Vulnerabilities, and Risks

Casually, what has been described above may appear to be nothing more than a threat/hazard assessment, vulnerability analysis, and risk assessment conducted in sequence. And there is indeed merit to doing so. In fact, institutions that regularly assess and evaluate threats, hazards, vulnerabilities, and risks are arguably more prepared for a variety of negative occurrences than those that prefer one over the other. Yet aligning these methodologies together provides a wider aperture to explore the linkages and connections across the entire biorisk spectrum.

Let's review the biorisk assessment process as laid out in the *Biosafety in Microbiological and Biomedical Laboratories* (BMBL) recently published in its sixth edition [6]. The focus of this six-step laboratory biological risk assessment methodology is on two major components: one, the biological agent in question, and two, the procedures by which the agent will be manipulated. In general, the steps include identification of characteristics of the biological agent, identification of laboratory process and procedure hazards, determining the appropriate biosafety level (e.g., BSL-2, BSL-3), review of biosafety controls with subject matter stakeholders, evaluation of worker proficiency with respect to control measures, and finally, adopting a process for changes to the risk assessment and control measures. This process is adaptable, scalable, and practical, and has been employed in many laboratory management programs in the U.S. and around the world. An analogous risk assessment process has been described by the World Health Organization [16]. The BMBL also includes a section of laboratory biosecurity and describes the major components of a biosecurity program that includes characterization of assets and potential threats, developing scenarios to identify likelihood and impact of occurrence, and a process for reevaluation [6].

While the sixth edition of the BMBL does inform both a biorisk assessment and biosecurity assessment paradigms, the connections between the biosafety and biosecurity aspects of each remain undefined. In other words, the continuity and directionality of a tethered threat, vulnerability, and risk assessment methodology is largely absent. This observation is not critical; in fact, it provides the users of the BMBL with great flexibility to tailor these guidelines to their individual operations. And establishing cross-linked, interoperable safety and security programs is challenging. So why are the connections, continuity, and directionality of cross-linked biosafety and biosecurity programs so important?

Consider an example where a potential risk has not been mitigated because the influences upstream, in the form of a threat or vulnerability, have not been identified to reveal the potential risk in the first place. How vulnerable is a vulnerability if an upstream threat has not been described? For example, a traditional biorisk assessment will adequately identify risks associated with an infectious agent that is aerosol transmissible in a procedure involving a centrifuge. But what threats, hazards, and vulnerabilities may be occurring upstream that would affect the likelihood or impact? In this case, a negligent laboratory worker (threat) with limited supervision (vulnerability) provides an excellent example. A biorisk assessment alone may not account for this combination of threat, vulnerability, and risk, and thereby not capture the increased probability of occurrence and impact. This is a small example of the importance of connecting these methodologies into a contiguous process.

To further support this, imagine working in a multi-faceted, sophisticated, large-scale environment with thousands of employees. Enterprises such as these require sizeable security operations and normally have an array of safety programs. What risks may be unknown to the head of laboratory safety if he/she is not aware of the threats, hazards, and vulnerabilities normally monitored by institutional security? What impacts to the institutional security programs may be unknown to those in charge because of lack of continuity with safety programs? These questions point out the critical interface that is necessary for integrated, interoperable biorisk management programs, inclusive of both biosafety and biosecurity concerns.

5 Summary

Taking the time to explore the many threats you face is critical to ensuring you are not the next headline. It is no longer reasonable to indicate "this will never happen here." Ask any of the companies that have suffered a loss due to an active shooter, theft of proprietary information, or a costly break-in. Threats and hazards come in all shapes and sizes and no organization is free of the risk that one of the threats they face will become a reality. At a certain point, your obligation to avert an otherwise recognizable risk will come front and center. Whether the loss is the result of the U.S. Occupational Health and Safety Administrations' (OSHA) interpretation of your obligation to provide a safe working environment after a domestic violence incident enters the workplace, or a loss of funding from your primary stakeholder in research due to publicized vulnerabilities and gaps that were exploited by others, your programs will be impacted significantly.

What this chapter has offered for the first time is a novel methodology that firmly establishes a contiguous process, effectively linking the biothreat and the biorisk assessment processes. We argue that addressing only one segment fails to provide insight into the meaningful connections that reside across the spectrum of threats, hazards, vulnerabilities, and risks. In a way, failure to adopt an upstream biothreat assessment process in conjunction with a downstream biorisk assessment process greatly limits the aperture of the individuals and institutions responsible for safeguarding people, animals, research, and property. We encourage institutions to consider the biothreat assessment process, combined with biorisk assessment methodologies, to develop more robust biorisk management programs.

References

1. ASIS International (2015) Risk assessment. ANSI/ASIS/RIMS RA.1-2015
2. ATAP (2016) Risk assessment guideline elements for violence: consideration for assessing the risk of future violent behavior. Association of Threat Assessment Professionals
3. Borum R, Fein R, Vossekuil B, Berglund J (1999) Threat assessment: defining an approach for evaluating risk of targeted violence. Behav Sci Law 17(3):323–337. https://doi.org/10.1002/(sici)1099-0798(199907/09)17:3<323::aid-bsl349>3.0.co;2-g. PMID: 10481132
4. Bretton-Gordon H (2020) Biosecurity in the wake of COVID-19: the urgent action needed. CTC Sentinel 13(11)
5. Burnette R (2013) Biosecurity: understanding, assessing, and preventing the threat. Wiley, Hoboken
6. Meechan PJ, Potts J (2020) Biosafety in microbiological and biomedical laboratories (6th edn). U.S. Dept. of Health and Human Services, Public Health Service, Centers for Disease Control and Prevention, National Institutes of Health, Washington, DC
7. Cronin AK (2003) Terrorist motivations for chemical and biological weapons use: placing the threat in context. Report for Congress. Congressional Research Service. https://fas.org/irp/crs/RL31831.pdf
8. FBI (2015) Making prevention a reality: identifying, assessing, and managing the threat of targeted attacks. U.S. Department of Justics, Federal Bureau of Investigation
9. Guldimann A, Meloy JR (2020) Assessing the threat of lone-actor terrorism: the reliability and validity of the TRAP-18. Forensische Psychiatrie, Psychologie, Kriminologie. 31 Mar 2020, pp. 1–9
10. Hernandez JC, Kovaleski SF (2009). Demanding job in a divided lab, then a murder. The New York Times. https://www.nytimes.com/2009/09/18/nyregion/18yale.html
11. Jackson BA, Frelinger DR (2007) Rifling through the terrorists' arsenal exploring groups' weapon choices and technology strategies. Studies in conflict and terrorist. Rand corporation. https://www.rand.org/content/dam/rand/pubs/working_papers/2007/RAND_WR533.pdf

12. Meloy R (2017) TRAP-18: Terrorist redicalization assessment protocol
13. Salerno RM, Gaudioso J (2015) Laboratory biorisk management: biosafety and biosecurity. CRC Press, Boca Raton
14. Security Vulnerability Assessment Methodology for the Petroleum and Petrochemical Industries (2003) American Petroleum Institute, Washington, DC, pp 3–28, Publication No. OS0002
15. United States, Department of Homeland Security, Office for Domestic Preparedness (2003) Vulnerability assessment methodologies report. Centralized scheduling and information desk, Washington, DC, pp 8–14
16. World Health Organization. Department of Epidemic and Pandemic Alert and Response (2004) Laboratory biosafety manual, 3rd edn. World Health Organization, Geneva

Expanding the Scope of Biosecurity Through One Health

Lauren Richardson

Abstract Biosecurity is an integral part of biorisk management in a laboratory, and its definitions and applications extend well beyond, impacting all aspects of human, animal, and environmental health. At a basic level, biosecurity aims to protect valuable biological material assets. Once the asset is well-characterized, vulnerabilities and potential threats can be assessed and managed in context of its nature and environment. Approaching threat assessment from the perspective of the biological asset —whether the asset in question is a pathogen, medicament, biotechnology, plant, animal, or associated data—offers opportunity to facilitate useful mitigations for its protection or conservation. Considering biosecurity through One Health's integrative, multi-disciplinary lens is key to achieve systemic health security for the humans, animals, and the environment. This approach requires critical evaluation and systems-level analysis, rather than limiting the scope of intervention to best practices. While the current speed of innovation threatens to outpace security, an understanding of the principles of biosecurity applied will facilitate decisions from the local to the global level.

1 Defining Biosecurity Outside of the Laboratory

Defining biosecurity outside of the context of a human laboratory can be confusing, as the term has a different meaning in the context of animal and plant agriculture. The World Health Organization defines laboratory biosecurity as "the protection, control and accountability for valuable biological materials… within laboratories, in order to prevent their unauthorized access, loss, theft, misuse, diversion or intentional release" [22]. The focus of biosecurity in the laboratory is intuitive: this is where agents are stored and studied, accessible in high concentration. The World Organization for Animal Health (OIE) Terrestrial Manual uses this definition, but does not describe its implementation in detail, generally addressing the concept with biosafety

L. Richardson (✉)
Merrick & Company, Washington, D.C., USA
e-mail: lauren.richardson@merrick.com

© Springer Nature Switzerland AG 2021
R. N. Burnette (ed.), *Applied Biosecurity: Global Health, Biodefense, and Developing Technologies*, Advanced Sciences and Technologies for Security Applications,
https://doi.org/10.1007/978-3-030-69464-7_3

[23]. This Manual encourages a risk assessment approach for comprehensive biorisk management: "biological risk assessments are undertaken to inform and determine the policy and procedures that in turn give confidence that the laboratory procedures for each of the biological materials handled by the laboratory pose negligible danger to a country's animal and human populations" [23]. As previously described, biosafety is directed to protection of humans from biological materials, and (in a threat-based approach) biosecurity focuses on the security of the valuable biological material within the laboratory as the target warranting protection. In this context, the execution of the OIE approach to biosecurity, which describes "an important biosecurity responsibility of veterinary laboratories and animal facilities to identify and to minimize any risk of release of pathogens into human and animal populations, either domestic or wild," focuses on biosafety.

Outside of the laboratory, biosecurity takes on different meaning, as it is defined in a broader context. Traditional agricultural biosecurity includes elements of biosafety, focusing broadly on infectious disease control and risk management. The Food and Agriculture Organization of the United Nations (FAO) defines biosecurity as "strategic and integrated approach to analyzing and managing relevant risks to human, animal and plant life and health and associated risks to the environment" [12]. The FAO Biosecurity Toolkit notes: "Biosecurity is a strategic and integrated approach to analyzing and managing relevant risks to human, animal and plant life and health and associated risks to the environment. Biosecurity covers food safety, zoonoses, the introduction of animal and plant diseases and pests, the introduction and release of living modified organisms (LMOs) and their products (e.g. genetically modified organisms or GMOs), and the introduction and management of invasive alien species. Thus biosecurity is a holistic concept of direct relevance to the sustainability of agriculture, and wide-ranging aspects of public health and protection of the environment, including biological diversity" [12]. This original use of the term "biosecurity" comes from the agricultural industry, where it describes measures to decrease infectious disease transmission in livestock and plant agriculture [16].

Though the term biosecurity has been used for decades in the context of agriculture and livestock, the focus has been on risk mitigation, rather than threat identification and mitigation. The American Biological Safety Association International (ABSA International) has recently moved to include the threat-based perspective, and to integrate the associated data security components, defining biosecurity as "the risk- and threat-based control measures established to prevent the unauthorized access, misuse, loss, theft, diversion and intentional release of valuable biological materials, pathogens, toxins, information, expertise, equipment, technology and intellectual property that have the potential to cause harm to humans, animals, plants, the environment, public safety or national security" [1]. Laboratory organizations may do well to apply lessons learned from the field, where impact of consequences forces action. Additionally, an understanding of systemic impacts at a global level may encourage similar expansion of biosecurity beyond agriculture to human health systems. Examining field biosecurity in the context of One Health from a threat management perspective is crucial to comprehensive threat and vulnerability analysis.

2   One Health Significance

One Health, "the integrative effort of multiple disciplines working locally, nationally, and globally to attain optimal health for people, animals, and the environment" [3], is an old concept with a new name, and its true reach has yet to be realized. Increasing ability to control one's environment, whether in the selection of an air-conditioned home or the purchase of a cheeseburger creates an artificial cognitive distancing between humans and their environment. The reality of increasing urbanization and globalization is different: urban areas encroach on forested areas, leading to unexpected interaction with wildlife, and spread of disease is only an airplane ride away. The interactions of humans and animals with one another and the built and natural environment are increasing—and becoming increasingly complex.

While the interplay between humans, animals, and their environments is easily demonstrated in smaller and more rural communities, where the impacts of these components on one another are easily seen, the value of an integrated approach has become an area of focus at a global level. Siloes of excellence within disciplines have facilitated development of specialized expertise, in some cases with an unfortunate effect of decreased understanding of impacts by other system components. While defined scope of agencies and organizations allows for measurement of project or program success and priority of funding, it often makes integration across sectors difficult. With renewed international focus on the One Health approaches, funding mechanisms for cross-cutting approaches may be made easier.

The interaction between human, animal, and environmental health is clear. Zoonotic infectious diseases pose significant risks to human health, and environmental conditions impact disease persistence and propagation. Sixty percent of human infectious diseases are zoonotic and seventy-five percent of emerging infectious diseases are of animal origin [25]. While potential for harm to human, agricultural, or environmental health can be brought by microbial agents not identified specifically as pathogens of security concern, the majority of agents listed as select within United States 42 CFR Part 73 affect animals, either as diseases of animal or zoonotic significance (see Fig. 1) [7]. Vectors required for disease transmission rely on specific environmental conditions—heat, moisture, airflow—to successfully complete their life cycles; this highlights the systemic nature of potential impact of pathogens. In light of recent advances in biotechnology and the increasing interaction of humans, animals and the environment, the potential for harm outpaces the ability to recognize potential threats. In addition to their role in disease transmission, propagation, and persistence, animal and environmental factors are useful in predicting and monitoring existing and emerging disease behavior. History of interaction with

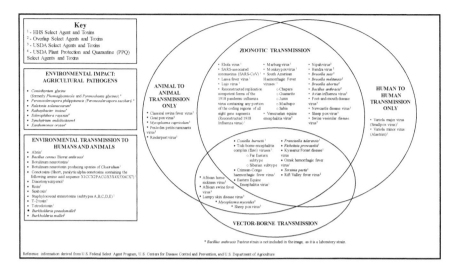

Fig. 1

animal or environmental reservoirs informs epidemiologic studies of human populations. For example, a history of abattoir work in an endemic anthrax area may aid in diagnosis of an individual human. Animal infectious disease outbreaks often precede human outbreaks, as the spillover of a natural animal outbreak reaches human populations. When combined with specific environmental conditions and animal practices, human disease outbreaks may be more predictable.

For example, Rift Valley fever (RVF) provides a clear example of One Health significance. RVF is a zoonotic disease that can cause illness in humans and livestock (i.e., sheep, goats, cattle, camels). The virus is transmitted via mosquito vector or through direct contact with fluids or tissues of infected animals. Though this disease is endemic in some areas, outbreaks occur following seasons of heavy rainfall, and eggs of infected animals may survice for years in the soil. Because of the predictable pattern of disease outbreaks, vaccination campaigns to prevent disease in animals may be initiated based on weather predictions [8, 17]. Understanding the disease process and ensuring collaboraton across sectors are crucial in disease control.

Food chain management, from production through distribution, is a potential area of interaction with human health systems. On a basic public health level, this is clear: food security is required to maintain optimal nutritional status for human health. Not to be confused with biosecurity, food security refers to access, availability, and utilization (ability to be used for nutrition) of food [26]. In terms of biosecurity, the food chain, with its numerous steps that involve domestic or international movement and exposure to many different individuals, offers many potential vulnerabilities for exploitation.

One Health is important in terms of not only health, but also economic stability. Animal and plant production touches agricultural development from smallholder farmers to international corporations, corporate and private sector production systems, intranational and international trade, environmental commodities (e.g., water systems), and health systems costs. In the public and sector, costs are realized in terms of direct healthcare services, economic losses to production, and trade exclusions. In the private sector, the cost may be felt most in loss of reputation or client base. Similarly, reputational costs outside of the health and agriculture sectors appear in losses due to decreases in dollars spent on tourism or other discretionary spending, as people lose confidence in security. Because human health is definitively linked to animal and environmental health, understanding of the integration and mutual dependencies between these areas of focus is useful in assessing threats and analyzing vulnerabilities.

3   Public Policy Context of Agricultural Biosecurity

Agricultural biosecurity is especially significant in the history of biosecurity guidelines and legislative development. The impact of threats to agricultural aspects are easily quantified in economic losses due to disease and subsequent compromise to food security. The ready visibility of this impact encourages engagement in assessment and mitigation to reduce threats in a way that may not be easily appreciated in a laboratory setting. Today, the majority of biosecurity legislation targets genetically modified organisms (GMOs): exclusion of GMOs is intended to protect corruption or competition with native wild and agricultural plant species. In 1993, New Zealand passed the first Biosecurity Act [5] so named, but the principles of agricultural biosecurity were well known, and associated legal restrictions were in place long before this law.

Several international organizations provide guidelines and recommendations for best practices in agricultural biosecurity. FAO recommends development of holistic biosecurity frameworks at the national, regional, and international levels, with needs and strategies determined through collaboration across sectors, focusing on integration and coordination of stakeholders [12, 13]. The FAO Toolkit offers a framework for development and implementation of national biosecurity programs, encouraging cross-sectoral collaboration and integration of stakeholders, and focusing on risk-based approaches [12].

Several analysis tools are in place to address food production systems security; however, these generally focus on food safety and hazard prevention and mitigation. Hazard Analysis Critical Control Point (HACCP) management systems facilitate analysis and control of hazards in food from raw material through finished product [21]. The World Trade Organization Agreement on Application of Sanitary and Phytosanitary [SPS] Measures allows countries to place scientifically-based restrictions on trade to protect animal, plant, and human health [29]. This agreement identifies the Codex Alimentarius, a collection of international food safety standards, funded by the World Health Organization (WHO) and FAO, as the organization setting these food safety standards [9]. The SPS agreement recognizes the OIE as the authority for animal health and zoonotic disease standards development.

In addition to these food systems guidelines, the FAO, OIE, and WHO work individually through the Performance of Veterinary Services (PVS) pathway [27] and International Health Regulations (IHR), as well as jointly through the Global Early Warning System (GLEWS) [14], to strengthen animal and human health systems to improve prevention, detection, and response to existing and emerging diseases. These organizations work to improve health systems, focusing on improving and coordinating surveillance and health access and response.

The existing legislation and frameworks demonstrate the value in targeting biosecurity concerns that are recognized at an international level. Agricultural biosecurity offers significant lessons learned that may be applied to strengthen biosecurity across health sectors and to a laboratory setting. At the national level, impacts to trade and tourism prompt action to preserve economic stability, and at the producer level, impacts to livestock health are easily quantified through economic losses. Here consequence informs priority. Taking care to quantify consequence is a lesson that could be expanded to the laboratory, where the risk/value proposition may facilitate higher priority for efforts to improve biosecurity.

4 Threat-Focused Biosecurity from a One Health Perspective

The Public Health Biosecurity Process is easily adapted to the field in a One Health Context. In *Biosecurity: Understanding, Assessing, and Preventing the Threat* [6], Dr. Burnette described the five pillars of security: physical, information, material, personnel, and transport. As we examine biosecurity in the context of One Health, the challenges extend beyond the laboratory environment to the field.

4.1 Physical Biosecurity

In assessing laboratory biosecurity, the physical aspects of security often receive the most attention: gates, guards, and guns are easy to understand and communicate with scientists, administrators, and security personnel. In the context of the laboratory, physical security is easily extended to understand the needs for workflow and role-based laboratory accesses. Animal holding facilities within the laboratory increase this complexity, as needs for animal restraint and holding facilities pose unique access challenges. As this extends outside of the laboratory, physical structures in animal housing facilities are rarely designed to protect the resources within from threats: designs often focus on protection from the elements, easy egress, and sufficient ventilation. Large-scale animal agricultural production facilities are often the exception to this pattern, with restricted entry and physical security for large poultry and livestock operations, in which introduction of pathogens would lead to significant economic consequences. At the leading edge of animal production biosecurity, commercial industry offers physical security restrictions that minimize potential contact with threats to the valuable biological material (the animals). These restrictions include building design and landscaping, entry access controls to the farm and animal housing, segregation by age and cohort, penning and gating strategies to minimize access and intermingling, and air handling mechanisms that reduce exposure to animals. The intent of such security is often designed to prevent natural infection; while the measures in place may offer similar protection, absence of threat assessment may result in gaps in physical security.

Outside of production and laboratory facilities, physical controls associated with animal housing are often minimal; an exception to this is in movement restrictions associated with compartmentalization[1] and zoning.[2] These designations are granted on a commodity[3] population or subpopulation and disease-specific basis by the OIE in recognition of physical separation of animal populations that may maintain disease from those that are free of disease. The intent is to define physical boundaries of subpopulations free from disease to encourage trading partners to allow importation of disease-free animals [2].

Physical boundaries are especially important for areas in which animals and humans live in close proximity, or those in which domestic animals have frequent contact with wildlife [28]. Pastoral societies, migrant populations, internally displaced persons, and refugees are at high risk for biosecurity threats. Application of the principles of physical security outside of the laboratory can be simple, but it warrants methodic examination of the process or locality, and feasibility may be limited due to cost or logistical constraints. Animals, humans, and environmental components often fail to recognize barriers, making physical security outside of a confined facility difficult. While disease status is often monitored geographically, even national borders may fail to protect one population from the next. Communities intermingle, making it difficult to secure naïve populations from pathogens brought by outside entities. Even change in land use and impacts to climate due to human activities increase the potential for zoonotic disease spread [10]. For example, deforestation has been linked to increases in zoonotic and vector borne disease transmission, with disruption of habitats and increased proximity between humans, animals, and vectors [28]. Understanding and predicting the behavior of the biological material (human or animal) may better inform public health entity attempts to offer this security.

Infectious disease events challenge any existing physical controls; outside of the laboratory, physical security controls are often costly and are generally limited in

---

[1]"Compartment means an animal subpopulation contained in one or more establishments under a common biosecurity management system with a distinct health status with respect to a specific disease or specific diseases [2]."

[2]"Zone/region means a clearly defined part of a territory containing an animal subpopulation with a distinct health status with respect to a specific disease for which required surveillance, control and biosecurity measures have been applied for the purpose of international trade [2]."

[3]"Commodity means live animals, products of animal origin, animal genetic material, biological products and pathological material [2]."

scope to address the issue at hand, and epidemics and emerging infectious diseases may require different measures. For example, a fence that prevents direct contact may not prevent vector transmission. In addition, stressors change the way animals and people interact with the environment, and changes in behavior may cause individuals to work around barriers. Resource diversion for control of disease may lead to weakening of boundaries, as priority moves elsewhere—this can inadvertently increase overall burden. In a corruption of triage principles, it is tempting to focus only on the urgent need of the moment. If a health system is overwhelmed by an epidemic, the base level of burden for other of diseases may be unchanged, and standard protocols must be maintained. Additional requirements emerge, and surge support may be unfamiliar with baseline requirements. During such events, a careful consideration of physical biosecurity controls should be included in any response activities, to ensure that both endemic and epidemic concerns are met appropriately.

4.2 Information Biosecurity

Even in the laboratory, it may be difficult to identify information that requires security, and the need for information security reaches further than currently realized. As aspects of information are digitized, the vulnerability to unwanted access may increase. With digitization, the volume of information continues to grow, and the ease of securing it becomes more complex. This combination of vulnerability and complex mitigation requirements is not limited to the laboratory.

Data associated with biological materials that warrants protection goes beyond individual personally identifiable information and health data, and an understanding of its potential value is key to recognizing the need for security. The intellectual property value and potential for misuse of data associated with biological materials in the laboratory are increasingly recognized as impetus for their protection, but information security outside of the laboratory may be more vulnerable and more valuable. Today's health technology landscape is extensive and potentially accessed by data owners and opportunists, stretching from private health data to databases maintaining epidemiologic information to simple devices that many do not recognize as Internet of Things (IoT). Increasing global connectivity provides increased opportunity and threat.

As the health and agricultural sectors become increasingly digitized, information security needs continue to grow. At the international level, traditional and non-traditional reporting channels and information sharing for disease incidence are often available through publically-available sources, and informal channels for communication (i.e. social media) are mined by both legitimate and malicious actors. At the national or sub-national and industry levels, health and agricultural data hold significant value. Public health data is useful for targeting specific populations. Proprietary information regarding pharmaceuticals and agricultural products is a ready target for corporate espionage or sabotage. Access to large volumes of data offer fuel to explore markets that drive research and production.

Ensuring information security outside of the laboratory requires coordination across sectors and staff. First, security professionals, information technology professionals, scientists, practitioners, and support staff must coordinate to understand the storage (e.g., databases, individual knowledge) and potential value of information. As highlighted during recent events, public interest in epidemiologic data and interpretation is growing, increasing its value, as the public recognizes the increased potential for impact on everyday lives. The COVID-19 pandemic has revealed gaps in information biosecurity, with threat actors seeking information associated with biological materials. For example, in 2020, the European Medicine Agency, which approves vaccines for the European Union, fell victim to cyberattack, with unlawful access of documents related to regulatory submission of COVID-19 vaccine candidates [4].

While not traditionally recognized as information security, the mechanisms for communication and context of release may also jeopardize the value of data. Individual or dissociated pieces of data may offer minimal value, but their aggregation or ability to be linked to individuals may create vulnerabilities for harm [4].

4.3 Material Biosecurity

Materiel security outside of the laboratory covers a broad range and varies by sector. Healthcare and agriculture sectors rely on a network of supply chain support from lab to pharmacy and from farm to fork. Pharmaceuticals have been a historic target for bad actors [11], and as personalized drug development and use of biologics rapidly expands, additional biosecurity measures will be required. This continues to be a serious concern: on 2 December 2020, Interpol released a global alert warning of potential targeting of COVID-19 vaccine supply chains [15]. Agricultural supply chains often span geographic areas and may be globalized, with a commodity sourced in one country, processed in another, and distributed in more. Traceability is vital for identification of potential vulnerabilities throughout this chain and for source identification in the event of an incident. Supply chain management, from procurement to distribution, varies widely among different field scenarios. The industrialized agricultural industry offers a framework for analysis. Because of the difficulty in resourcing in many locations with limited access to traditional supply resources, effecting consistent resource streams may not be possible. Traceability and chain-of-custody principles can be adapted to biological materials used in field scenarios. Livestock traceability is internationally recognized as an important component of food safety. Individual identification requirements at the global level may be less stringent, but animal traceability is easily linked to health. While logistically and ethically challenging, an increased focus on linking health data to individuals in a way that may be readily accessed is worthy of reflection. In the COVID-19 pandemic, the importance of contact tracing has received significant attention, with digital applications emerging to support a tracking of individual health status. This may be a worthwhile endeavor, but it deserves scrutiny in its execution and potential implications. Waste is not often considered valuable biological material, but it may be a readily-available source of infectious material for use by nefarious actors.

Waste management often receives dangerously insufficient attention in sectors outside of the laboratory. In high containment laboratories, a single discarded sample is treated with care to ensure decontamination or destruction is complete; in homes, farms, and medical facilities, especially in less-developed areas, waste management may not be given the same attention. Even unprocessed samples may hold value, as they may offer resources to facilitate development of control measures. In 2016, WHO organized a review to support the characterization, preservation, and use of biological samples taken during the 2014-2015 Ebola outbreak in West Africa [24]. In outbreak scenarios, the volume of waste may become overwhelming: institution appropriate, scalable biosecurity controls to address waste management is key. Waste generated from biological incidents, such as the Ebola virus disease epidemic in West Africa in 2014–2015, demonstrated the importance, and gaps, of material biosecurity in infectious disease scenarios requiring surge support (chapter "Applied Biosecurity in the Face of Epidemics and Pandemics: The COVID-19 Pandemic").

Material security consideration is required from origin to disposal; for this reason, traceability of commodities and products supports transparency in assessment. Understanding waste management and potential sources, disposal, and access is crucial to ensure proper access, accounting, and treatment of potentially dangerous material.

4.4 Transport Biosecurity

Concerns related to material and transport security are often intertwined when addressing activities outside of the laboratory. Movement controls outside of the laboratory are much more difficult to apply and enforce, as the authority for intervention and resources available are much more limited. Human movement controls may rely on individual recognizance in response to health professional recommendations; in some circumstances, integration with law enforcement may be required.

Sample movement from the field to the laboratory may be difficult to execute, even without security considerations. For example, in resource-challenged areas, couriers often have limited access to transport, and may rely on shared or public transportation. In remote areas, access to transport throughways may be limited seasonally or throughout the year. In unstable or high crime areas, couriers may be at risk for carjacking; biological materials in transit are at risk of inadvertent theft. Sample containers may not be secured or appropriately temperature controlled.

Monitoring of animal movement is potentially more complicated than that of human movement. Traceability requirements for animal livestock vary and are often tied to trade; many systems lack requirements for traceability in animals that do not cross boundaries or are not involved in international trade. As previously discussed, domestic animals often interact with wildlife, whose movement may be entirely uncontrolled. Pastoral societies may not recognize regional or national boundaries, and animal movement is difficult to track. Even in well-resourced nations, animal movement may be difficult to monitor, as controls and reporting requirements are often difficult to enforce. While identification of individual humans in transit across national borders is tightly controlled, it is far from simple. Restrictions on human movement face legal, ethical, and feasibility limitations. As seen during the recent COVID-19 pandemic, movement restrictions and stay-at-home orders can be difficult to enforce, especially in low resource areas and where legal requirements and recommendations of adjacent jurisdictions differ. With minimal economic support to those in perilous economic positions, an individual's own needs often outweigh their desires to participate in behaviors that protect their communities. Competing risks and threats may be prioritized by community leaders, who may lack access to resources to offer relief to their community members. This is not to say that one risk or threat is inherently of greater priority than another; they are often intertwined. For example, following closure of meat procession plants due to COVID-19 spread in worker populations, farmers were forced to depopulate herds, leading to economic losses in increased wholesale product cost [17, 18]. Risks and threats exist in a One Health context, with significant downstream effects of any mitigation strategy; cross-cutting evaluation and implementation of mitigations is needed to avoid unintended consequences.

4.5 Personnel Biosecurity

Personnel security is difficult to capture adequately, regardless of the circumstance, as any attempt to assess security risk is a snapshot in time. In a laboratory environment, background checks, training, and culture that emphasizes security facilitate development of good personnel security programs. In a healthcare or field scenario, this becomes difficult. Personnel are often recruited on-site to assist in care or handling in support of a project: animal handlers are best suited to provide care for general safety, and families and community health workers are most likely to have access. Many do not have a background in healthcare or science; here the balance of efficiency and security does not fit into an easily applicable, repeatable framework. When including personnel without a scientific or healthcare background, training in basic biosafety and biosecurity is vital and should be adapted to the task and the individuals.

In addition to personnel screening and training needs, there are behavioral tendencies that help and hinder efforts outside of the laboratory, especially in those who work with animals. An awareness of these behaviors can be helpful in designing programs and projects that incorporate civilians. Those outside of the scientific fields may also be more likely to place trust in those within their own community, whether local or through social media, than in science professionals. This is an extension of the "culture of safety and security" that many laboratories seek to develop, and the tendency for mimicry of leadership will always outperform training or education. This is exacerbated when the value proposition of safe, secure behavior is not obvious.

Clear risk and threat communication are vital and should be backed by an underlying explanation of reasoning for mitigation steps, which have been validated for not only logical value, but also feasibility. This is easily demonstrated in most agricultural settings, as the economic impact is measurable. For example, animal handlers may be reluctant to use PPE and appropriate techniques for biosecurity, due to lack of access to supplies, perceived or actual impediment to activities, or lack or concern about potential for exposure. An explanation of the economic value—reduction of disease transmission between animals as a result of decreased propagation by and through humans—coupled with consistent, obvious practice by personnel in positions of trust and authority will be more effective than a lecture on the values of biosecurity accompanied by a "do as I say, not as I do" approach to practice.

With an increase in both data and access to social platforms and a decrease in perceived need for security, the risk for intentional or unintentional insider threat is high. Perceived or real pressure to engage in information sharing may be exacerbated in times of stress. Who speaks matters: individuals may be better equipped to communicate with those in their own communities. An understanding and appreciation of the needs for biosecurity and the potential economic repercussions of biosecurity failure means that support from those involved is more forthcoming.

5   Lessons Learned

It is easy to segregate animal biosecurity from laboratory biosecurity, viewing them as related but separate goals in risk reduction; however they serve ultimately similar purposes in the intent to protect biological material.

There are several lessons to be learned from current efforts.

- The traditional definition of agricultural biosecurity can be extended to include aspects of laboratory biosecurity. Understanding a systems perspective allows risk and threat assessors to identify vulnerabilities and areas for intervention. Individuals assessing biosecurity can incorporate industry-specific systems assessments, identifying practices of interest.
- When assessing biosecurity from a One Health perspective, it is important to invite all the appropriate players during assessment. At the facility or program level, this includes participants from each group of staff, as well as representative civilians who may be involved with a project. At the national level, this means including representatives across sectors (e.g., agriculture, health, finance, education, law enforcement), as well as relevant industry and citizen sectors.
- Understanding a framework is important. Disease-based and species-specific interventions may be necessary but are insufficient alone.
- Impediments for incorporation of biosecurity measures may be unrelated to scientific understanding.
- Absence of stable governance, or presence of excessive or cumbersome bureaucratic requirements may limit efforts in coordination and cooperation.

- Insufficient funding may limit interventions or result in inappropriate prioritization of efforts. With financial resources, biosecurity is often not a priority. Parties with more funding are often permitted to make decisions for implementation.
- Social and cultural beliefs and concerns may limit the reach of interventions.

Laboratories appear to be closed systems, tightly controlled so that biosecurity is almost an afterthought, a checklist to complete. An appreciation of the process of threat assessment, mitigation, and communication used in agricultural biosecurity clarifies that this is untrue and confirms its distinction from the realm of biosafety. Protection of biological materials, whether in a vial or on a field, exists in a dynamic environment and has systemic impacts to individual communities and in a global health context. Recent events have increased visibility into the health and life sciences industries, and the ability for institutions to offer transparency in their diligence and methodical nature has become a key component of risk management.

6   Summary

With increasing biotechnology development and globalization, the importance of the bioeconomy—"economic activity that is driven by research and innovation in the life sciences and biotechnology, and that is enabled by technological advances in engineering and in computing and information sciences [20]" —has drawn significant attention. Human, animal, and environmental health are often considered independently, but in truth they are inseparable: holistic approaches are implicit in protection of the bioeconomy. Apparently dissociated workstreams coalesce and dramatically impact one another. Technology that improves production in one area may simply displace risk. Thoughtful, collaborative approaches are the way forward. The One Health context demonstrates the unique challenges to different sectors and solutions that can be achieved through a thorough understanding of the intrinsic nature and environmental context of the biological material. Additionally, lessons learned in the laboratory or the field may be leveraged to resolve challenges in both areas.

As personnel with limited formal education and foundational scientific knowledge increase their participation in complex endeavors, the critical thought and careful consideration underlying practices is often not included in what people "know." This superficial understanding of the practices may be dangerous, as individuals work around the processes designed to keep them safe. An N95 mask offers protection only if chosen appropriately, fitted properly, and worn correctly. In this context, improving knowledge and underscoring critical thought in risk and threat assessment offers an adaptable approach to safe, secure science and health in and outside of the laboratory.

The One Health framework is a useful standalone concept, but it serves best as a holistic, cross-cutting lens to view actions. Several sectors warrant additional exploration to support One Health efforts in biosecurity. Agriculture, animal health, clinical human healthcare, biotechnology and biomanufacturing, environment and natural resources, and infrastructure programs would benefit from this shift in perspective, to encourage exploration of applied biosecurity concepts. As public health, laboratory institutions, and agricultural systems become a more frequent, prominent focus on the public stage, transparency and security must strike a balance. Science is becoming more egalitarian, with increasing expectations and capabilities and decreasing barriers to entry that present growth of opportunities and threats. A One Health approach, encompassing systemic risk, threat, and vulnerability management will be vital in the new landscape of innovation.

References

1. ABSA International (2020) 1st biosecurity virtual symposium. ABSA International. https://biosecuritysymposium.org/#:~:text=Biological%20security%2C%20biosecurity%2C%20is%20the,and%20intellectual%20property%20that%20have
2. World Organisation for Animal Health (2011). Terrestrial animal health code, vol. I, 20th Edn
3. American Veterinary Medical Association (2018) One HealthWhat is One Health? https://www.avma.org/KB/Resources/Reference/Pages/One-Health94.aspx
4. BioNTech (2020) Statement regarding cyber attack on European medicines agency https://investors.biontech.de/news-releases/news-release-details/statement-regarding-cyber-attack-european-medicines-agency
5. New Zealand Ministry for Primary Industries (2020). Biosecurity Act 1993. Retrieved from https://www.legislation.govt.nz/act/public/1993/0095/latest/DLM314623.html
6. Burnette R (2013) Biosecurity: understanding, assessing, and preventing the threat. Wiley, Hoboken
7. CFR Title 42 Part 73—Select Agents and Toxins, (2005).
8. Centers for Disease Control and Prevention (2016) The story of the rift valley fever virus vaccine. https://www.cdc.gov/onehealth/in-action/rvf-vaccine.html
9. Food and Agriculture Organization of the United Nations & World Health Organization (2018). Codex Alimentarius International Food Standards. Retrieved from http://www.fao.org/faowho-codexalimentarius/en/
10. Environmental Resilience Institute (2020) COVID-19 and environmental change: The social and environmental factors contributing to the rise of zoonotic diseases. The Trestees of Indiana University. https://eri.iu.edu/tools-and-resources/fact-sheets/zoonotic-diseases.html
11. Fletcher D (2009) A brief history of the tylenol poisonings. TIME. https://content.time.com/time/nation/article/0,8599,1878063,00.html
12. Food and Agriculture Organization of the United Nations (2007) FAO biosecurity toolkit. FAO. https://www.fao.org/3/a1140e/a1140e00.htm
13. Food and Agriculture Organization of the United Nations Committee on Agriculture (2003) Biosecurity in food and agriculture, item 9 of the provisional agenda, 31 March4 April 2003
14. GLEWS+ (2020) The joint FAO-OIE-WHO global early warning system for health threats and emerging risks at the human-animal-ecosystems interface. https://www.glews.net/
15. INTERPOL (2020) INTERPOL warns of organized crime threat to COVID-19 vaccines. https://www.interpol.int/en/News-and-Events/News/2020/INTERPOL-warns-of-organized-crime-threat-to-COVID-19-vaccines
16. Meyerson LA, Reaser JK (2002) A unified definition of biosecurity. Science 295(5552):44. https://doi.org/10.1126/science.295.5552.44a
17. Mintert J (2020) COVID-19 impacts on pork and beef. https://ag.purdue.edu/commercialag/home/resource/2020/04/covid-19-impacts-on-pork-and-beef/
18. National Pork Producers Council (2020) Hog farmers face COVID-19 financial crisis. https://nppc.org/hog-farmers-face-covid-19-financial-crisis/

19. Terrestrial Animal Health Code, Volume I, Twentieth Edition (2011). https://www.oie.int/doc/ged/D10905.PDF
20. The National Academies of Sciences, Engineering, Medicine (2020). Retrieved from https://www.nationalacademies.org/news/2020/01/us-bioeconomy-is-strong-but-faces-challengesexpanded-efforts-in-coordination-talent-security-and-fundamental-research-are-needed
21. U.S. Food and Drug Association (1997) HACCP principles & application guidelines. https://www.fda.gov/food/hazard-analysis-critical-control-point-haccp/haccp-principles-application-guidelines
22. World Health Organization (2006) Biorisk management: laboratory biosecurity guidance. https://www.who.int/csr/resources/publications/biosafety/WHO_CDS_EPR_2006_6.pdf
23. World Organisation for Animal Health (2017) Manual of diagnostic tests and vaccines for terrestrial animals 2017, 7 edn, vol. 1–3. https://www.oie.int/standard-setting/terrestrial-manual/access-online/
24. World Health Organization (2020) Biobanking Ebola samples. WHO. https://www.who.int/blueprint/priority-diseases/key-action/biobanking_ebola_samples/en/
25. World Organisation for Animal Health (2013) The 'One health' concept: the OIE approach. *OIE News Bull.*
26. World Food Programme (2018) What is food security? https://www.wfp.org/node/359289
27. World Organisation for Animal Health (2018) The OIE PVS Pathway. https://www.oie.int/en/support-to-oie-members/pvs-pathway/
28. Zimmer K (2019) Deforestation is leading to more infectious diseases in humans. National Geographic. https://www.nationalgeographic.com/science/2019/11/deforestation-leading-to-more-infectious-diseases-in-humans/
29. World Trade Organization (1995). Agreement on the Application of Sanitary and Phytosanitary Measures

Biodefense Promotes Biosecurity Through Threat Reduction Programs and Global Health Security

Brittany Linkous, Ryan N. Burnette, and Samantha Dittrich

Abstract As the world faces an increasing rate of emergent and re-emergent infectious disease events, whole-of-government biosecurity systems are critical now more than ever for countries to detect and contain biothreats at their source. While most disease outbreaks occur naturally, it is important to prepare for potentially devasting outbreaks caused by an intentional release of a dangerous agent. Threat reduction programs and global health security aim to prevent the unauthorized access, loss, theft, deliberate release, or misuse of hazardous biological agents and associated-related information and actively promote responsible conduct of life science research and oversight of dual-use risks. Laboratory staff who have access to especially dangerous pathogens can avert intentional releases through appropriate training, tools, and oversight in biosecurity. Ultimately, biosecurity practices help countries to counter natural and manmade biological threats while also fostering safe scientific progress. Strong biosecurity capacity is key to building defenses and optimizing global health security against biological threats. This chapter will focus on the application of biosecurity in major U.S. and international biodefense and threat reduction programs, as well as analyze the pivotal role of biosecurity in the context of global health security.

1 Global Health, Threat Reduction, and Biodefense Initiatives Are Intertwined

In our interconnected world, an infectious disease threat anywhere can present as a threat everywhere. Past decades have seen several new pathogens emerge and old pathogens resurge. Severe acute respiratory syndrome (SARS), avian influenza H5N1, Zika, and now the 2019 novel coronavirus (COVID-19) are recent examples of new and re-emerging pathogens that have had tremendous impacts on human, animal, and economic health across the globe. Future threats are likely to arise as

B. Linkous · R. N. Burnette (✉) · S. Dittrich
Merrick & Company, Washington, D.C., USA
e-mail: ryan.burnette@merrick.com

© Springer Nature Switzerland AG 2021
R. N. Burnette (ed.), *Applied Biosecurity: Global Health, Biodefense, and Developing Technologies*, Advanced Sciences and Technologies for Security Applications,
https://doi.org/10.1007/978-3-030-69464-7_4

the global population continues to grow, the demand for food becomes greater, and microbes become increasingly resistant to treatments such as antibiotics. The need to be prepared to prevent, detect, and respond to threats posed by nature, bioterrorism, and the accidental release of agents from a laboratory is more apparent than ever before.

An infectious disease outbreak anywhere in the world has the potential to become an epidemic, or worse, a pandemic, as the 2019 novel coronavirus (COVID-19) is showing the world all too well. New bacteria and viruses are emerging, and others are growing resistant to existing treatments. With the increase in travel and trade, a pathogen can travel worldwide to major cities in as little as 36 hours. Moreover, evidence has shown how easily and quickly epidemics can derail economies. Experts estimate that the 2003 SARS outbreak cost the global economy between $30 and $40 billion in just six months. The economic impact of the ongoing COVID-19 pandemic is unprecedented in scope and magnitude. The United Nations (UN) Department of Economic and Social Affairs estimates $8.5 trillion in cumulative losses during 2020 and 2021 due to lockdowns, the closing of national borders, and job losses. Further, an estimated 34.3 million people will likely fall below the extreme poverty line in 2020, with 56% of the increase expected in African countries [16].

In 2005, member states of the World Health Organization (WHO) agreed to revise the International Health Regulations (IHR 2005), a legal framework requiring countries to develop a minimum level of capacity to detect, assess, and report public health events. The regulations identify eight core capacities including national legislation policy and financing, surveillance, response, preparedness, risk communication, human resources, and laboratory. The purpose of which is to prevent, protect against, control, and provide a public health response to the international spread of disease. Under the IHR, all countries must report events of international public health importance. The regulations were revised to address new and emerging epidemic threats directly, including, smallpox, poliomyelitis due to wild-type poliovirus, human influenza caused by a new subtype and SARS, cholera, pneumonic plague, yellow fever, viral hemorrhagic fever, West Nile fever, and other biological, radiological, or chemical events that meet IHR criteria [1]. By 2007, 169 countries had committed to reaching compliance with the regulations by the year 2012 [13]. More than two-thirds of countries failed to meet these standards by 2014 [10]. Moreover, there were no common targets or external assessment mechanisms in place to help countries quickly identify and fill gaps.

Given the ongoing threat posed by emerging infectious diseases, attention to global health security—efforts supporting epidemic and pandemic preparedness and capabilities—has grown tremendously over the past few decades and even more so in 2019 amid the COVID-19 pandemic. Although the U.S. government has engaged in efforts to address global health security for more than two decades, its involvement has expanded in recent years, and global health security is now a defined component of the U.S. global health response. Through its threat reduction programs such as the Cooperative Threat Reduction (CTR) Biosecurity Engagement Program (BEP) and leadership, launch and ongoing activities of the Global Health Security Agenda (GHSA), the U.S. has helped other countries make measurable improvements in their capacities to prevent, detect, and response to public health threats. A critical function of these efforts, threat reduction programs, and the GHSA is advancing biosecurity.

2 Department of State (DOS) Bureau of International Security and Nonproliferation, Office of Cooperative Threat Reduction (ISN/CTR) Biosecurity Engagement Program (BEP)

With the increase of transnational crime and terrorism, emerging and re-emerging infectious diseases, biotechnology capabilities and techniques, in combination with the varying levels of biosafety and biosecurity practices employed have significantly altered how biological threats are viewed not only in the U.S., but throughout the world. In an effort to mitigate these threats, the U.S. National Security Council evaluated strategies for strengthening pathogen security across the globe, identifying proliferation issues in critical geographic regions. Consequential to these findings, the addition of the CTR BEP was tasked with combating emerging biological threats, enhancing threat awareness capabilities, developing prevention and protection tool, enhancing disease surveillance and detection, and developing post outbreak or incident response and recovery capabilities in the priorities regions that were identified [11]. In essence, BEP's mission has extended the spirit of the NunnLugar engagement, with a broadened global scope.

2.1 History of the U.S. Department of State Biosecurity Engagement Program

Up until 2006, ISN/CTR's budget was largely focused on engagement activities with former Soviet Union scientists, with support specifically focused towards science and technology centers. However, due to the increase in potential biological threats in other regions of the world, CTR BEP's mission pivoted in 2009 to focus on preventing terrorists and proliferators from acquiring weapons of mass destruction (WMDs) expertise, materials and technology and expand programs in locations where terrorist threat is highest [11].

The U.S. Department of State (DOS) nonproliferation programs encompass a variety of activities to meet the mission of preventing biological threats. Since 2006, these activities include, but are not limited to dismantlement, physical security of facilities, laboratory biosecurity, physical infrastructure upgrades, salaries and equipment for researchers, biosafety, regulatory training, disease surveillance, public health, and one health. BEP's inaugural objective was to promote legitimate bioscience research while recognizing the confluence of bioterrorism threats, emerging infectious diseases, and the rapid expansion of biotechnology; focusing on areas in South Asia, Southeast Asia, and the Middle East. Over the past 12 years, BEP has expanded their objectives to also include dual-use science, global health security, One Health, veterinary biosafety and biosecurity, risk assessments, physical security of laboratories, and biorisk management.

BEP's nonproliferation mission up until recent years focused on the three pillars of biosecurity:

(1) **Increase biosafety and biosecurity** through technical consultations, risk assessments, and training courses, and build the human capacity and internal expertise to create a sustainable culture of laboratory biorisk management;

(2) **Strengthen the capacity** for public and veterinary health systems to detect, report, and control infectious disease outbreaks; and,

(3) **Enhance global health security** and foster safe, secure, and sustainable bioscience capacity through joint scientific collaborations designed to help prevent, detect, and respond to biological threats [2].

Cross cutting through each of these three biosecurity pillars, BEP increasingly focused on the incorporation of long-term sustainability in an effort to build capacities and infrastructure to ensure the impacts of the activities continue after U.S. funding has concluded.

However, engagement funding opportunities as recently as 2020, BEP's pillars have once again shifted to remain dynamic and reactive to the global needs within the biosecurity arena. As of 2020, BEP's national security mission is, "to mitigate global biological threats by minimizing the access of proliferator states and non-state actors to biological expertise, materials (e.g., high consequence pathogens [HCP]) [18]. The two pillars that BEP is currently aligning programmatic efforts towards are the following:

(1) **Denying Non-State Actors** the Expertise, Materials, and Equipment Necessary to Conduct Biological Attacks, and Preventing and Mitigating Accidental and Natural Outbreaks of High Consequence Pathogens;

(2) **Thwarting State Actors** from Developing and/or Advancing BW Efforts by Safeguarding Advanced Bioscience Research Facilities and Preventing Access to BW-Applicable Knowledge, Materials, and Equipment [18].

The current pillars (as of 2020), align BEP's programs and mission toward the advancement of biosecurity and nonproliferation efforts and goals while reacting to current environments, threat trends and needs across the countries and regions of engagement. In recent years there has been an increase in the number and types of non-state actors threats, including but not limited to terrorists, insider threats, and science hobbyists (e.g. do-it-yourself biologists [DIY-Bio]). While the pillars utilize more threat focused language than seen in previous years, BEP's national security mission continues to support the long standing value of sustainable scientific engagement and capacity building.

In order to achieve their nonproliferation mission, BEP employs a variety of methods that are specific to the region, culture, and activity objective. These methods include, but are not limited to: technical consultation and risk assessments, training

Table 1 Countries that have received support through ISN/CTR BEP funded activities

Region	Partner country
Central and South America	Brazil, Mexico, and Peru
Sub-Saharan Africa	Guinea, Kenya, Uganda, Nigeria, South Africa, Somalia, Mali, Sierra Leone, Senegal, Liberia
Middle East and North Africa	Iraq, Jordan, Saudi Arabia, United Arab Emirates, Turkey, Lebanon, Yemen, Egypt, Morocco, Libya, Algeria, and Tunisia
South Asia	Pakistan, Afghanistan, and India
Southeast Asia	Indonesia, Philippines, Malaysia, Thailand, Cambodia, Vietnam, and Bangladesh
Eurasia/Eastern Europe/Central Asia	Azerbaijan, Georgia, Russia, Ukraine, Kazakhstan, Kyrgyzstan, Tajikistan, Turkmenistan, and Uzbekistan

programs on biorisk management, train-the-trainer programs, safety and security upgrades, support of biosafety associations, surveillance and capacity building development and support, disease diagnosis, adoption and compliance of international frameworks and regulations, field epidemiology and laboratory training, research collaborations, training support and grants. Furthermore, BEP's mission success is a result of their extensive network of subject matter experts (SMEs) and implementers, who facilitate a large portion of BEP's activities on their behalf through cooperative agreements. By utilizing these resources, BEP has been able to focus on increasing biosecurity efforts, securing life science institutions, preventing non-state or state actor acquisition of HCPs, while developing projects that meet each partner country's specific needs and priorities.

Since its initial engagements in Eurasia, the Former Soviet Union, Indonesia, and Iraq, BEP has expanded its focus to countries in South Asia, the Middle East, North and Sub-Saharan Africa. Up until 2020, BEP's highest priority countries for engagement were Iraq, Turkey, and Yemen, followed by their second highest priority countries, Afghanistan, Algeria, Egypt, India, Indonesia, Jordan, Lebanon, Libya, Malaysia, Nigeria, Philippines, Somalia, and Tunisia. While BEP does not identify specific countries of priority in their most recent portfolio, BEP does indicate that their priorities remain in Southeast Asia, South and Central Asia, Sub-Saharan Africa, Near East, Europe, and Eurasia [18]. The table below shows all the countries that have been in partnerships or engagements with BEP since 2006 (Table 1).

2.2 Biosafety Association Support and Partnerships

In an effort to increase country partnership sustainability and inclusion of their needs and priorities as they relate to biosafety, biosecurity, and public health, BEP has supported, partnered, or assisted in the development of biological safety associations. These associations include, but are not limited to: the American Biological Safety

Association (ABSA International), the International Federation of Biosafety Associations (IFBA), Asia-Pacific Biosafety Association (A-PBA), European Biosafety Association (EBSA), African Biological Safety Association (AfBSA), Philippines Biosafety & Biosecurity Association (PhBBA), Brazil Biosafety Association (ANBIO), Biosafety Association of Pakistan (BSAP), Biosafety Association of Mexico (AMexBio), and the Indonesian Biosafety Association (IBA).

Through the development, support, and partnership with international biosafety associations, BEP is able to sustain in-country relationships, while promoting biosafety and biosecurity best practices after the initial engagement has ended. Moreover, by promoting these organizations biosafety and biosecurity professionals from across the globe are incorporated into an international network where they are able to share best practices, lessons learned, and develop new partnerships outside of the BEP funding framework. Biosafety associations have developed into a necessary component of advancing BEP's mission and creating sustainable biorisk management programs in partner nations. It is likely these associations will remain a viable point of partnership for the U.S. Department of State, as well as other international entities.

A Future in International Cooperative Engagement and Sustainability

Over the past several fiscal years, BEP's portfolios have received fluctuating levels of funding. In FY17, funding allocations were at a high, with funding ceilings up to $40 million for the fiscal year [17]. However, in more recent years, funding allocations that have been made available for implementers noted a dramatic drop in funding, as the FY19 notice of funding opportunity (NOFO) declared only $5 million available for program implementation [18]. In the FY21 NOFO, available funding rose to $14 million to support foreign assistance activities related to BEP programming. As these funding trends continue in the wake of growing government concern of terrorism and the potential acquisition of biological agents and materials, BEP is likely to focus their support and engagement on activities that enhance biosecurity of biological facilities, address concerns regarding emerging technologies, as well as insider threat programs. As a result, BEP will need to engage and continue to engage with other stakeholders in the international community, such as the United Kingdom Biological Engagement Program (U.K. BEP) and the Canadian Global Partnership Program (GPP) to ensure efforts are consistent and not duplicative.

While BEP remains a successful biological nonproliferation and engagement program as other U.S. and international programs engage the same regions, cooperation and information sharing will become essential to ensure repetitive activities are minimized, the successes and lessons learned from each program is capitalized on, and the impact of each program remains top priority. Additionally, this will provide the partner countries with the most value for their engagement and efforts, as well as promote forward momentum of their biological programs to ensure they are understanding and compliant with international standards,[1] can operate their programs

---

[1]BEP specifically supports and promotes the adoption of and compliance with comprehensive international frameworks that advance U.S. biological nonproliferation objectives, including United

effectively, and have the capability to become self-sustainable post-engagement. Furthermore, with the future funding of U.S. biological engagement programs left to congressional appropriation, in conjunction with the changes in administrations and fluctuating U.S. foreign policy, an international cooperative approach will become essential to sustaining long-term results. In general, engagement remains a cost-effective model to disseminate and advance capacities and capabilities associated with biosafety and biosecurity. The U.S. Department of State is but one of many international agencies and ministries that have adopted a model of cooperative engagement.

3 U.S. Department of Defense: Defense Threat Reduction Agency

The U.S. Department of Defense (DoD) plays a significant role in all aspects of defending the U.S. from foreign attacks of any kind. This includes biodefense as a major objective. The Defense Threat Reduction Agency (DTRA) was formed in 1998 following the consolidation of several DoD entities, although the origins of today's DTRA could be traced back to the Armed Forces Special Weapons project as early as 1947. The Agency's next major evolution was catalyzed by the fall of the Soviet empire, thought to possess thousands of nuclear weapons, thousands of tons of chemical weapons, and an aggressive biological weapons program. Today, DTRA houses the Cooperative Threat Reduction Directorate which collaborates with partner nations to "secure, eliminate, detect, and interdict WMD-related systems and materials." This Directorate accounts for three major programs: (1) Proliferation Prevention Program, (2) Security and Elimination Programs, and (3) Biological Threat Reduction Program (BTRP).

3.1 History of the Biological Threat Reduction Program

BTRP (formerly Cooperative Biological Threat Reduction) is a comprehensive program that seeks to partner with nations toward a combined goal of reducing the threat of biological agents being used as weapons. The origins of BTRP can be traced back to biological weapons elimination following the collapse of the former Soviet Union [6]. The DTRA model of engaging partner nations to offer financing and expertise has resulted in a large portfolio of projects and activities aimed as reducing the threat of biological agents as weapons. More recently, BTRP has expanded to include some overlap in mission space with the Global Health Security Agenda. DTRA was

Nations Security Council Resolution (UNSCR) 1540, the Biological Weapons Convention (BWC), the World Health Organization's International Health Regulations, and the European Committee for Standardization (CEN)/International Organization for Standardization (ISO) standards.

largely confined to the former Soviet Union (FSU) until the U.S. Congress authorized expansion into other geographies in 2008.

This ushered in a period of growth in both geographies and capabilities. The primary mission had always been to consolidate materials that pose a threat, support partner nations to secure those items, and subsequently reduce the number or eliminate them altogether. In many respects, those historical mission areas remain relevant today. Some of the lasting impacts of the BTRP efforts are most visible in the FSU, which included significant support to laboratory infrastructure and capacity building.

3.2   DTRA's Biological Threat Reduction Program Today

While the past 25 years saw major efforts in the FSU, in the form of laboratory construction, renovation, upgrades, and capacity building, today's BTRP includes a host of other program areas all aligned with reducing the threat of biological agents. Notably, these programs now include biosurveillance, biorisk management program development, and even funding support for threat reduction research efforts in partner countries. It is these newer areas of BTRP that begin to overlap with the Global Health Security Agenda, recognizing that not all biological threats are manmade. Rather, threat reduction must also account for naturally occurring biological threats as well. Today, DTRA has partnerships with countries in the FSU, Africa, the Middle East, and Southeast Asia.

4   Biodefense Programs Around the World: Promoting Biosecurity

Clearly biodefense programs are reliant upon more than just threat reduction, such as those with the DoS and the DoD/DTRA. Today, most developed nations have some form of a biodefense program that accounts for biological threat reduction towards human, animal, or ecological assets. Most of these programs are summarized in various national-level strategies or plans, through international agreements, and multi-sectoral partnerships.

Perhaps the most prominent of the multi-national organizations anchored to biodefense is the Biological Weapons Convention (BWC) under the United Nations Office for Disarmament Affairs. The BWC was the first "*multilateral disarmament treaty banning the development, production, and stockpiling of an entire category of weapons of mass destruction…*" and was formally instituted in 1975 [15]. Today, the BWC has 183 states parties. However, one major and persistent criticism of the BWC is that is lacks authority to monitor its parties for compliance. Rather, the BWC encourages collaboration and transparency.

4.1 National Level Biodefense Strategies Reflect Biosecurity Principles

U.S. President Trump issued the National Biodefense Strategy in 2018 with five major goals that align with (1) risk awareness and risk-based decision making; (2) augment national-level capabilities to deter, detect, deny, and prevent natural or manmade bioincidents; (3) reduce the impacts of bioincidents; (4) ensure capabilities to respond rapidly to bioincidents; and (5) develop actions to support recovery efforts following a bioincident. Despite being a national-level strategy, concepts are anchored in biosecurity rhetoric, including "deter, detect, deny," and "response and recovery."

Similarly, in 2018 the U.K. released the U.K. Biological Security Strategy. Like the U.S., the U.K. established a governance board that allows for collaboration across a variety of stakeholder ministries. The goals described in the U.K.'s plan are more closely aligned with the Global Health Security Agenda, by way of (1) understanding the nature of the biological risks, (2) preventing the entry of biological risks, (3) achieving suitable capabilities to detect and (4) respond to biological risks. Measures such as these in the U.S. and the U.K. support the reality that biodefense, and health security, are not limited to any one department or agency; rather, multi-sectoral collaboration, information sharing, and resourcing are necessary.

Similar national-level plans, strategies, and policies are found in most developed nations. What is notable is that a major component of biodefense strategies is the augmentation of biosecurity programs. As described by threat reduction programs, a major strategy to build a partnership network for biodefense is augmenting individual country's biosecurity capabilities. Biodefense initiatives are supported by biosecurity program augmentation in the following ways:

Consolidation of biological agents and materials that have weapons potential: Great efforts have been expended in many developed countries to consolidate dangerous pathogens that have the ability to create health and economic damage. In theory, this reduces the number of stockpiles of biological materials in unaccounted for locations. Consolidation also implies that greater stringency on inventory management, access control, and even regulations permitting access, can take place.

Reduction or elimination of biological stores with weapons potential: DTRA's work in the FSU is a prime example of this aspect, and the premise is simple- destroy biological materials to prevent them from falling into the hands of malicious actors [6]. The goal of this element is largely obvious as elimination of biological agents of threat reduces the overall threat.

Building capabilities and capacities for safe, secure handling and storage: A major push from many international partners, such as the U.S., Netherlands, Germany, France, U.K., and others has been to work cooperatively with partner nations to build national-level policies, training programs, increasing diagnostics, biosurveillance, disease reporting, upgrading laboratory function and security, and establishment of subject matter expert associations to augment biorisk management practices. In theory, this provides those nations with greater capabilities to consolidate, reduce/eliminate, and secure those biological agents that may pose a threat.

The primary point is that much of what constitutes a biodefense initiative or program is the promotion of national and local biosecurity practices and principles. This "grass roots" approach is intended to put greater control and accountability for potentially dangerous biological agents in the hands of practitioners. And yet, much of this is managed in developed nations in the absence of prescriptive legislation [3]. As we will explore further in this chapter, many of these principles fall into the Global Health Security Agenda as well.

4.2   Biodefense Includes Agricultural Defense

The coronavirus pandemic has demonstrated the tremendous economic and human health costs. At the time of this writing, the toll is still being accounted for, but the losses on all fronts are nothing short of devastating. It stands to reason that protecting human life is always a priority but we cannot forget the level to which human health is integrated with animal, plant, and environmental health as well (see chapter three "Expanding the Scope of Biosecurity Through One Health)." Further, entire economies are built upon agriculture and trade, tightly woven into a global marketplace. The U.S. Department of Agriculture (USDA) reported that agriculture, food, and related industries contributed over $1 trillion to U.S. gross domestic product and accounted for almost 11% of employment in the country in 2019 [19]. Therefore, protecting agriculture and the food supply has great importance.

The USDA is primarily responsible for protecting the security of the agricultural and food sectors in the U.S. with significant collaboration from the U.S. Department of Homeland Security, the U.S. Department of Justice, and the U.S. Food and Drug Administration. A notable example and historical institution is the Plum Island Animal Disease Center (PIADC). Located off the tip of Long Island, NY, PIADC has been the primary laboratory defense installation protecting U.S. livestock from foreign animal disease. It also remains the only entity in the U.S. legally possessing and studying foot-and-mouth disease. The laboratory has been in operation since 1954 and is early in the process of closing its doors making way for a new facility, the National Bio and Agro-defense Facility in Manhattan, KS. Similar institutions operate around the world to ensure the security of agriculture and the food supply, including The Pirbright Institute in the U.K., the National Animal Health Laboratory in New Zealand, the Australian Centre for Disease Preparedness, the Canadian Science Centre for Human and Animal Health, and others. Many of these laboratories represent the leaders of branch laboratories throughout their respective countries that provide sentinel diagnostics and collaborative research, forming a network of stakeholder laboratories.

5 The Global Health Security Agenda

Launched by the U.S. Government (USG) in February 2014, the Global Health Security Agenda (GHSA) is an effort by nations, international organizations, and civil society to "build countries' capacity to help create a world safe and secure from infectious disease threats and elevate global health security as a national and global priority" [8]. The main goal of the GHSA is to strengthen both global and national capacity to prevent, detect, and respond to human and animal infectious disease threats through a multi-lateral, multi-sectoral approach. This underlying Prevent-Detect-Respond framework is guided by multi-sectoral and international coordination and communication, rapid and effective response, prevention and reduction of outbreaks, and early detection of emerging threats.

5.1 Overview

When the GHSA initiative started, the U.S. committed to assisting 31 countries and the Caribbean Community in achieving 11 measurable GHSA targets. The U.S. is investing $1 billion in resources across 17 "Phase I" countries to build capacity to prevent, detect, and respond to future infectious disease outbreaks (Table 2). This funding has helped these countries establish a five-year country roadmap to achieve and sustain each of the targets of the GHSA. These roadmaps are intended to capture national priorities for health security, bring key stakeholders and sectors together, identify partners, and allocate resources for strengthening health security capacity. The U.S. is also working with 14 additional "Phase II" GHSA partner countries and the Caribbean Community (Table 2) to finalize GHSA roadmaps and mobilize international partner resources to reaching the GHSA targets [4]. In 2016, President Obama issued an executive order to launch a comprehensive framework for the GHSA. This established a senior-level policy coordination mechanism within the U.S. for the GHSA initiative and defined specific roles and responsibilities for participating agencies. The Executive Order also advanced the U.S.'s capacity to partner with new sectors to address epidemic threats and leverage leadership to synchronize assistance across health systems [20].

Since its launch, the GHSA has become a global collaborative, multi-sectoral, and multi-stakeholder initiative that aims to accelerate and optimize global health security. Through a partnership of more than 70 countries, international organizations, and non-governmental stakeholders, the vision driving the GHSA is a world safe and secure from global health threats posed by infectious diseases. The GHSA leverages and complements the strengths and resources of global partners to build and improve country capacity and leadership to prevent, detect, and respond effectively to infectious disease threats. Over the years, the GHSA established an approach that enhances and encourages coordination, active engagement, and multisectorality for the country and non-country members. This mutual commitment relies on actions

Table 2 Phase 1 and 2
GHSA countries

Phase 1 GHSA countries	Phase 2 GHSA countries
Bangladesh	Cambodia
Liberia	Laos
Burkina Faso	Democratic Republic of Congo
Mali	Malaysia
Cameroon	Mozambique
Pakistan	Georgia
Côte d'Ivoire	Peru
Senegal	Ghana
Guinea	Rwanda
Sierra Leone	Haiti
Ethiopia	Thailand
Tanzania	Jordan
India	Ukraine
Uganda	Kazakhstan
Indonesia	CARICOM*
Vietnam	Laos
Kenya	Malaysia

*Caribbean Community (CARICOM) is an organization of 15
Caribbean nation and dependencies

of key stakeholders to emphasize global health security as a national-level priority
and galvanize tangible commitments to undergo planning and resource mobilization to address gaps. This includes sharing best practices, elevating global health
security as a priority of national leadership, and facilitating national capacities to
comply with the IHR, the World Organization of Animal Health (OIE) Performance
of Veterinary Services (PVS) pathway, and other relevant international health security standards, frameworks, and strategies. The partnership is led and supported by
a Steering Group comprised of approximately 15 countries, international organizations, and/or non-governmental stakeholders, including the GHSA Consortium
and GHSA Private Sector Roundtable. The main role of the Steering Group is
to provide strategic guidance and direction, including identifying overall GHSA
priorities, tracking of progress and commitments, and facilitation of target-driven
multi-sectoral coordination and communication among the GHSA members [8].

5.2   The GHSA Action Packages

In 2014, member countries identified eleven discrete GHSA Action Packages to
facilitate both regional and global collaboration towards specific GHSA objectives

and targets. Each action area is based on the IHRs and has leading and contributing member countries and includes a five-year target, an indicator (or indicators) for measuring progress, and a baseline assessment, planning, monitoring, and evaluation activities to support successful implementation. All GHSA countries have flexibility in how they address their commitments, with countries participating in one more of the action areas and working nationally, regionally, and/or globally toward the common targets. All activities and multi-sectoral initiatives within the GHSA contribute to preparedness and health systems strengthening with the common goal to enable countries to prevent, detect, and respond to any health emergency risk. The purpose of the Action Packages and the underlying Prevent-Detect-Respond framework is to accelerate measurable and coordinated actions in support of GHSA, highlight targets, and provide accountability for countries to not only make specific commitments but also to take leadership roles in the GHSA effort.

The Action Packages detail practices for preemptively guarding against threats to global health security, establish mechanisms by which a country can determine when a threat has arisen, and emphasize the tools and capacity necessary to address threats as they are occurring, and include goals to establish emergency operations centers. These specific targets in the Action Packages include countering AMR; preventing the emergence and spread of zoonotic diseases; advancing countries' biosafety and biosecurity systems; enhancing immunization; establishing national laboratory systems; strengthening surveillance; strengthening disease reporting; supporting workforce development; establishing emergency operations centers; linking public health law, and multi-sectoral response; and enhancing medical countermeasures and personnel deployment (Table 3).

Table 3 The global health security agenda is comprised of 11 action packages	11 GHSA action packages
	Antimicrobial resistance
	Biosafety and biosecurity
	Emergency operation centers
	Immunization
	Laboratory systems
	Medical countermeasures and personnel deployment
	Public health and law enforcement
	Surveillance
	Sustainable finance
	Workforce development
	Zoonotic disease

5.3 The Joint External Evaluations

Working closely with the WHO, the U.S., Finland, and other GHSA Steering Group representatives supported the development and implementation of the Joint External Evaluation (JEE). The JEE is a voluntary, collaborative external assessment tool to identify gaps in capacity, determine a baseline, and measure progress in a country's ability to prevent, detect, and respond to infectious disease and other health threats. The JEE combines GHSA Action Packages with additional capabilities required under the IHR, allowing countries to identify the most urgent needs and address specific gaps within their health security system. The tool specifically helps to highlight gaps and needs for current and prospective donors and stakeholders and to inform national priority setting, target resources, and track progress. As of November 2020, 110 countries completed their JEE, and 12 JEEs were in the pipeline [23].

In terms of the JEE process, the evaluation is completed in two stages. During the first stage, the country conducts an initial self-evaluation using the JEE tool and works to complete a self-evaluation report in collaboration with relevant in-country representatives and stakeholders. This provides a self-reflection of the country's IHR capacities and capabilities across 19 technical areas and also includes all GHSA-related capacities. After the self-evaluation is completed, the host country shares it with the JEE Team comprising of experts from member states, WHO, OIE, FAO, Interpol, and other key international organizations. The host country experts then present their country's capacity in the technical areas covered in the JEE tool over several days of presentations and discussions with all relevant sectors.

During the second stage, an external evaluation team of subject matter experts conducts an in-country evaluation in close collaboration with the country. The JEE Team assigns scores for each indicator as well as identifies strengths and best practices, areas that need strengthening, challenges, and three to five key priority actions for each technical area. The JEE Team travels to the host country, enhancing their knowledge and understanding of the host country's capacities and capabilities. Once complete, preliminary results are presented to the country's high-level representatives from across the relevant sectors, and then a final report is written within two weeks after the mission is finished. This final report is shared with the country for feedback and then posted online. Countries are then able to utilize the data and lessons learned from the evaluation process in order to inform country-level planning and priority setting, including the development of a national action plan for health security (GHSA roadmap) to achieve targets (Fig. 1). Follow-up evaluations are recommended on a regular basis to track progress toward the achievement of the JEE targets and full IHR and PVS compliance [23].

Fig. 1 Notional GHSA "Roadmap" as a potential means to achieve country-specific targets

5.4 GHSA 2024

Since its inception in 2014, the GHSA has accelerated political and multisectoral support for health security and catalyzed a level of national and international action towards preventing, detecting, and responding to infectious disease threats. Through shared commitment, participation, transparency, and rigor, many countries, international organizations, and non-governmental and private sector entities have made tangible progress in strengthening critical core capacities. In a statement cementing GHSA as a national priority, former CDC Director Dr. Tom Frieden noted that in the first two years of the GHSA, "an outbreak of avian influenza was rapidly recognized and stopped [in Cameroon], with the emergency team in place within 48 hours compared with a response to past emergencies that took eight weeks or more to organize. In Uganda, outbreaks of cholera, meningitis, and Yellow Fever have been rapidly identified and stopped. And in Tanzania, GHSA enabled a more rapid and more effective response to cholera" [7].

During the Fourth Annual GHSA Ministerial in Uganda in October 2017, Ministers issued the Kampala Declaration to continue work to strengthen global health security and extending the GHSA until 2024. This next phase of GHSA, GHSA 2024, aims to be more strategic and streamlined, advancing a multi-sectoral approach, supporting adherence to international human and animal health standards, and advancing sustainable financing for global health security efforts across all sectors. In this next phase, Action Package working groups are continuing to focus on the original 11 action packages from the first phase of GHSA. However, they will work to advance eight Action Package Working Groups as part of the Prevent-Detect-Respond Framework (Table 4). To further support the strategic objectives of the GHSA, the Steering Group established flexible, time-limited Task Forces to advance priorities in a targeted way and ensure this work leverages and complements the efforts of partners and other health security actors. Bringing together interested GHSA members, the priority areas of the initial Task Forces include multi-sectoral stakeholder engagement, accountability and results, and advocacy and communications. There is also one permanent Task Force, the Action Package Working Group Coordination Task Force, that is responsible for updating the Steering Group on Action Package activities and progress and ensuring the Action Packages remain focused on furthering GHSA's primary objectives and core principles.

Table 4 The 2024 GHSA contains eight of the eleven action packages for advancement

8 GHSA 2024 action packages
Antimicrobial resistance
Biosafety and biosecurity
Immunization
Laboratory systems
Surveillance
Sustainable finance
Workforce development
Zoonotic disease

GHSA 2024 offers an opportunity to further harmonize the GHSA with one health strategies; strengthen pandemic preparedness of and communities' access to the global health workforce; address emerging health and biosecurity threats of international concern; reduce systemic barriers; and maximize the role of all relevant stakeholders, including civil society and the private sector. Over this 5-year mandate, GHSA, along with relevant partners, will actively contribute to national, regional, and global efforts to support countries in evaluating, planning, mobilizing resources, and implementing of activities that build health security capacity. Its goal is that by 2024, more than 100 countries (1) will have completed a joint external evaluation of health security capacity, (2) undergone planning and resource mobilization to address gaps, and (3) will be in the process of implementing activities to achieve impact. These countries will strengthen and demonstrate improvements in at least five technical areas to a level of 'Demonstrated Capacity,' as measured by relevant health security assessments, such as the WHO IHR Monitoring and Evaluation Framework [9].

5.5   Biosafety and Biosecurity Action Package: A Closer Look

Within the GHSA, biosafety and biosecurity represent a foundational and fundamental action package to meet the GHSA's mission and goals. The GHSA Action Package on Biosafety and Biosecurity (APP3) aims to advance global biosafety and biosecurity by reducing the risk of both deliberate and accidental spread of dangerous pathogens to human and animal populations through a whole-of-government approach. The target of APP3 is for countries to complete a national framework and to have a comprehensive oversight system for pathogen biosafety and biosecurity, strain collections, containment laboratories, and monitoring systems. APP3 uses two indicators to measure a country's biosafety and biosecurity capacity: (1) whether a whole-of-government biosafety and biosecurity system is in place for human, animal, and agriculture facilities and (2) whether biosafety and biosecurity training and practices exist [8]. Employing biosafety and biosecurity practices ensures that pathogens are identified, secured, and monitored appropriately.

APP3 provides synergistic opportunities with other international initiatives that have common biosafety and biosecurity capacity building priorities and objectives such as the IHR (2005), the Biological Weapons Convention (1975), and the United National Security Council Resolution (2004). What is unique about APP3 is that although APP3 focuses exclusively on biosafety and biosecurity, components of laboratory biosafety and biosecurity capacities are cross-cutting across the GHSA action packages. For example, the Action Package Detect 1 focuses on national laboratory systems, which could affect biosafety and biosecurity standards. A potential downside noted in the current GHSA framework is the apparent limitation of biosecurity to the laboratory level. It is possible that future GHSA action package refinement may lead to an expanded role of biosecurity in application.

Since the implementation of the first phase of the GHSA, partner countries have made significant progress in establishing or strengthening multi-sectoral systems capable of responding effectively to global health threats. Yet, the role of biosafety and biosecurity is often deprioritized and overlooked. Released in 2019, the Global Health Security Index found that 66% of countries lacked adequate policies for biosafety, 81% of countries lacked adequate policies for biosecurity, and only 1% of countries had appropriate oversight for potential dual-use life science research with especially dangerous pathogens [12]. This analysis reveals that most countries have limited core biosafety and biosecurity capabilities and subsequently, a lack of preparedness for preventing, detecting, and responding to bioterrorism threats. Even countries with advanced laboratory systems and public health capacity are scoring low on the biosafety and biosecurity metrics. These scores not only indicate a lack of capability in areas critical to countering biological threats, but they also underscore the need for continued and coordinated assistance across sectors to improve biosecurity and biosafety and to meet the target and objectives of APP3 under the GHSA.

Given that biosafety and biosecurity intersect with many other areas of the action packages and processes throughout the Prevent-Detect-Respond framework, this highlights the need to think in a more progressive manner and start applying security measures outside the laboratory premise as discussed in the next section. This may include access control particularly in isolation areas and emergency response center; personnel authorization; integration of biosecurity training modules in healthcare workers training; control measures applied throughout the lifecycle of a sample or specimen and any potentially contaminated item; and awareness and ownership of biosecurity measures by staff members operating throughout health systems.

6 Novel Biosecurity Applications and Threat Reduction Programs

6.1 The Role of the Laboratory

Laboratories are an integral component of public health systems and play a critical role in the detection, prevention, and control of diseases. National laboratory systems ensure wide geographic coverage for disease surveillance, adequate delivery of services and evidence-based laboratory data to inform public health security measures, particularly for national borders. They also strengthen biosafety and biosecurity, support efforts to counter drug resistance and support early warning of the emergence of new diseases.

As demonstrated by the recent Ebola and Zika virus outbreaks and the ongoing COVID-19 pandemic, it is essential that public health laboratories have the capacity to work safely with high consequence and emerging pathogens and to provide appropriate guidance to clinical laboratories. As such, the role of national health laboratories in support of public health response has expanded beyond laboratory testing to include a number of other core functions such as emergency response, training and outreach, communications, laboratory-based surveillance, and data management.

When looking closely into the response to diseases outbreaks, it is easy to see how security risks begin early on in the disease process with the collection of blood and other body fluids at clinics and emergency operation centers. Samples can be stored at these locations for hours or even days. They are often transported between clinics and laboratories, a step which most often happens on roads that lack the physical and transportation security measures required such as sample handling, storage, and disposal in laboratories. Therefore, biosafety and biosecurity are integral components of laboratories.

At the laboratory level, biosafety and biosecurity are sets of principles and practices aimed, in part, at containing infectious biological agents. Biosecurity is often used to refer to mechanisms to establish and maintain the security and oversight of pathogenic microorganisms, toxins, and relevant resources. However, international organizations and agreements use the word biosecurity in a variety of contexts and for different purposes. The WHO defines laboratory biosecurity as the "institutional and personal security measures designed to prevent the loss, theft, misuse, diversion, or intentional release of pathogens and toxins" [21]. This is accomplished by limiting access to facilities, research materials and information. FAO and OIE refer to biosecurity in the context of biological and environmental risks associated with food and agriculture. These risks include everything from the introduction and release of genetically modified organisms (GMOs), the introduction and spread of invasive alien species, animal diseases, and zoonoses to the erosion of biodiversity, the spread of transboundary cattle diseases, and the preservation of food supplies after production [22].

Laboratory biosecurity employs principles and practices aimed at identifying and mitigating threats and corresponding vulnerabilities that may result in the loss, theft, misuse, or diversion of biological assets throughout the disease process [5, 22]. Robust laboratory biosecurity programs are built on the "five pillars," described as physical security, personnel reliability, material control and accountability, transportation, and information security [3, 14]. These five pillars are commonly codified through laboratory site-specific policies, procedures, operations, training, and physical/information infrastructure. Further, the resulting laboratory biosecurity program is integrated into broader laboratory management and quality systems. Such laboratory programs help prevent unwanted access to sensitive biological agents and the associated information in addition to assist in response and recovery efforts in the event of a loss of containment.

Laboratory biorisk management is fundamentally a culture of rigorously assessing risks posed by working with infectious agents and toxins in laboratories and deciding how to mitigate those risks. It includes a range of practices and procedures to ensure the biosecurity, biosafety, and biocontainment of those infectious agents and toxins. Countries must be prepared to prevent, detect, and respond to threats posed by bioterrorism and the accidental release of viruses from a laboratory. For these reasons and others, the role of the laboratory, armed with robust biorisk management systems, are foundational to global health security and threat reduction efforts.

6.2 Biosecurity Outside of the Laboratory

While biosafety and biosecurity measures have traditionally been focused and applied in laboratory settings, public and global health experiences with multiple outbreaks have shown the need to take and apply biosecurity measures outside the walls of laboratories. Dangerous pathogens are not just confined to laboratory vials and storage tubes. In many instances, they originate in communities and are handled in clinics and isolation facilities that are far from considered safe or secure, particularly in emergency response settings. There is also the risk of the potential use of laboratories to make and release, whether intentionally or not, dangerous microbes. Genetic-engineering tools, unfortunately, have made it easier for terrorist groups or lone madmen to unleash custom-designed killer bioweapons.

In order to be prepared to prevent, detect, and respond to threats posed by bioterrorism and accidental release of viruses from a laboratory, biosecurity needs to be integrated into biorisk management. Laboratory biorisk management is fundamentally a culture of rigorously assessing risks posed by working with infectious agents and toxins in laboratories and deciding how to mitigate those risks. It includes a range of practices and procedures to ensure the biosecurity, biosafety, and biocontainment of those infectious agents and toxins not just at the laboratory level but in the field [22]. Although biosecurity cannot change the intent of malicious individuals who get their hands-on biological agents, applying best practices of how to prevent and respond to outbreaks can, in fact, minimize access.

Biosecurity is more than just the safeguarding of dangerous pathogens from individuals or organizations who would use them for harm. It is taking into consideration all current and future biorisks, and presenting ways to identify, prevent, and minimize them.

7   Summary

Biosecurity is a fundamental component of programs that support human health, animal health, and environmental health and is indispensable to both threat reduction programs and global health security. Although threat reduction programs and the GHSA aim to advance biosecurity, there is still an urgent need to prioritize biosecurity as a public health priority. Much remains to be done to strengthen biosecurity, including support for policy changes, legislation development, capacity building, training, education, infrastructure, management, analysis, information sharing, and outreach. While disease outbreaks are typically natural occurrences, an outbreak could be caused by an accidental release of dangerous pathogens or intentional dissemination of lethal agents by those intending to cause harm. Diseases know no boundaries and disease outbreaks can devastate the health of populations, cost billions in economic losses, destabilize political environments, and threaten national security.

Global health security means safer nations, more stable economies, and fewer failed states. Preventing the next outbreak and accidental or intentional release of dangerous pathogens will continue to require close collaboration between the health, animal, agriculture, defense, security, development, and other sectors. This will require close cooperation of defense, public health, animal health, and biorisk management practitioners. Countries will need to continue to employ an interconnected global network that can respond effectively to limit the spread of infectious disease outbreaks in humans and animals, mitigate human suffering and the loss of human life, and reduce economic impact.

References

1. Baker MG, Fidler DP (2006) Global public health surveillance under new international health regulations. Emerg Infect Dis 12(7):1058
2. Biosecurity Engagement Program (2016) About US. https://www.bepstate.net/about-us/locations/
3. Burnette R (2013) Biosecurity: understanding, assessing, and preventing the threat. Wiley, Hoboken
4. Centers for Disease Control and Prevention (2018) Advancing the Global Health Security Agenda: CDC Achievements and Impact—2017. https://www.cdc.gov/globalhealth/security/ghsareport/images/ghsa-report-2017.pdf
5. Centers for Disease Control and Prevention and National Institutes of Health (2004) Biosafety in Microbiological and Biomedical Laboratories, 4th edn. www.cdc.gov/biosafety/publications/bmbl5/BMBL.pdf

6. DTRA (2015) The cooperative biological engagement program research strategic plan: addressing biological threat reduction through research. U.S. Department of Defense: Defense Threat Reduction Agency and U.S. Strategic Command Center for Combatting Weapons of Mass Destruction

7. Frieden T (2016) President Obama Cements Global Health Security Agenda as a National Priority. https://blogs.cdc.gov/global/2016/11/04/president-obama-cements-global-health-sec urity-agenda-as-a-national-priority/

8. GHSA (n.d.) Key messages from the Global Health Security Agenda action package prevent-3: Biosafety & Biosecurity. https://ghsagenda.org/wp-content/uploads/2020/07/ghsa_2pager_final_print-app3.pdf

9. GHSA (2018) GHSA 2024 overarching framework. https://ghsagenda.org/

10. Gostin LO, Katz R (2016) The International Health Regulations: the governing framework for global health security. Milbank Q 94(2):264–313

11. National Research Council (2007) The biological threat reduction program of the department of defense: from foreign assistance to sustainable partnerships. National Academies Press, Washington, pp 43–56

12. Nuclear Threat Initiative and Johns Hopkins Center for Health Security (2019) Global Health Security Index. https://www.ghsindex.org/wp-content/uploads/2019/10/2019-Global-Health-Security-Index.pdf

13. Rodier G, Greenspan AL, Hughes JM, Heymann DL (2007) Global public health security. Emerg Infect Dis 13(10):1447

14. Salerno RM, Gaudioso J (2015) Laboratory Biorisk Management: biosafety and biosecurity. CRC Press, New York

15. United Nations (n.d.) Biological Weapons. United Nations Office for Disarmament Affairs. Retrieved from: https://www.un.org/disarmament/wmd/bio/

16. UN Department of Economic and Social Affairs (2020) World Economic Situation and Prospects as of mid-2020. https://www.un.org/development/desa/dpad/wp-content/uploads/sites/45/publication/WESP2020_MYU_Forecast-sheet.pdf

17. U.S. Department of State Bureau of International Security and Nonproliferation, Office of Cooperative Threat Reduction (ISN/CTR) (2017) Global Biosecurity Engagement Activities. U.S. Department of State Bureau of International Security and Nonproliferation, Office of Cooperative Threat Reduction (ISN/CTR) Notice of Funding Opportunity. Grants.gov. https://www.grants.gov/web/grants/view-opportunity.html?oppId=29861

18. U.S. Department of State Bureau of International Security and Nonproliferation, Office of Cooperative Threat Reduction (ISN/CTR) (2020) FY21 Global Biosecurity Engagement Activities. U.S. Department of State Bureau of International Security and Nonproliferation, Office of Cooperative Threat Reduction (ISN/CTR) Notice of Funding Opportunity. Grants.gov. https://www.grants.gov/web/grants/search-grants.html?keywords=SFOP0007376

19. USDA (2020) Ag and Food Sectors and the Economy. USDA Economic Research Service. Retrieved from: https://www.ers.usda.gov/data-products/ag-and-food-statistics-charting-the-essentials/ag-and-food-sectors-and-the-economy/#:~:text=Agriculture%2C%20food%2C%20and%20related%20industries,about%200.6%20percent%20of%20GDP

20. The White House, Office of the Press Secretary (2016) Statement by National Security Advisor Susan E. Rice on the Executive Order on Advancing the Global Health Security Agenda. https://obamawhitehouse.archives.gov/the-press-office/2016/11/04/statement-national-security-advisor-susan-e-rice-executive-order

21. WHO (2004) Laboratory Biosafety Manual, 3rd edn. https://www.who.int/csr/resources/publications/biosafety/Biosafety7.pdf

22. WHO (2006) Biorisk Management: Laboratory Biosecurity Guidance. https://www.who.int/ihr/publications/WHO_CDS_EPR_2006_6.pdf?ua=1

23. WHO (2020) Strategic Partnership for International Health Regulations (2005) and Health Security (SPH). https://extranet.who.int/sph/

Applied Biosecurity in the Face of Epidemics and Pandemics: The COVID-19 Pandemic

Samantha Dittrich, Lauren Richardson, and Ryan N. Burnette

Abstract Biosecurity as a discipline remains largely defined by institutional-level practices unbounded by the guidelines and checklists prevalent in the field of biosafety. The result is a high-level of interpretation at the individual and institutional level of defining and implementing biosecurity. At its core, biosecurity frameworks are largely anchored to a mirrored process of risk assessment and management. Therefore, biosecurity is still rooted in the fields of threat and vulnerability assessment, analysis, and management. The aperture of threat and vulnerability management may at first seem contrary to the overall field of public health where emerging infectious disease, and the negative consequences, are more commonly perceived as risks. However, in the spectrum of threat to vulnerability to risk (see Chaps. "Redefining Biosecurity by Application in Global Health, Biodefense, and Developing Technologies" and "The Biothreat Assessment as a Foundation for Biosecurity") infectious disease as an entity emerges as a threat. The ability or inability to respond and defend from infectious disease can be characterized as vulnerabilities. The negative consequences (e.g., spread of, infection, mortality, economic impacts) exist as a probability of occurrence, or simply, a risk. This is increasingly evident at the time of composition of this chapter and this book during the global COVID-19 pandemic. This chapter explores the parallels and distinctions of biosecurity-related concepts as they apply to the COVID-19 pandemic, lessons learned from previous epidemics and pandemics, and offers suggestions of stronger connectivity to threat and vulnerability management concepts as we inevitably prepare for future epidemics and pandemics.

S. Dittrich · L. Richardson · R. N. Burnette (✉)
Merrick & Company, Washington, D.C., USA
e-mail: ryan.burnette@merrick.com

© Springer Nature Switzerland AG 2021
R. N. Burnette (ed.), *Applied Biosecurity: Global Health, Biodefense, and Developing Technologies*, Advanced Sciences and Technologies for Security Applications, https://doi.org/10.1007/978-3-030-69464-7_5

1 Biosecurity is an Applied Discipline in Epidemics and Pandemics

To date, the diversity of definitions and nebulous landscape have allowed the field of biosecurity to avoid the checklist-heavy mentality that has become inherent to the field of biosafety. The intensity of regulation and adoption of best practices into minimum requirements has encouraged a movement of biosafety from a macro-scale of risk assessment to a micro assessment of the individual laboratory spaces. This has served well the high containment laboratories with varied and constant risks to personnel and dedicated biosafety staff to ensure procedures and practices are appropriately developed and maintained, but it may have fewer desirable impacts to laboratories with lower levels of containment and more constrained resources. As evidenced by the COVID-19 pandemic, the need for a risk and threat-based approach to applied science is vital. Biosecurity offers a lens through which to view such global concerns outside of the laboratory or production facility.

Though the majority of media attention has focused on epidemiologic metrics and public health practices to prevent infectious disease spread, the cross-cutting nature of biosecurity allows an integrated view of these disciplines. The principles of biosecurity are key in applying concepts associated with biostatistics, laboratory diagnostics, and individual pathogen behaviour to identify and mitigate risks, reduce vulnerabilities, and address threats to shape outcomes in a pandemic scenario.

What is less clear is the discrete role that biosecurity plays in the detection, response, and recovery efforts of epidemics and pandemics, scenarios often defined by containment measures, infection control, infection prevention, hygiene, and public safety. Perhaps the difficulty is assigning a discrete role for biosecurity is the overall homogeneity of these interrelated disciplines. Clearly biosecurity practices alone are not sufficient to prevent or respond adequately to an event as widespread and catastrophic as the current COVID-19 pandemic. Here, we attempt to identify those discrete points, their overlaps with related principles and practices, and suggest a meaningful method for employing biosecurity (e.g., biothreat/biohazard management) principles to such a global event. As with many technical areas, it is often beneficial to review similar events from our past.

2 Lessons Learned from Previous Epidemics and Pandemics

Although the COVID-19 pandemic is unprecedented, many lessons can be drawn from previous outbreaks. The proliferation of clinical and diagnostic laboratories in the developing world has resulted in a greater capacity to detect and diagnose infectious diseases. However, underdeveloped biosecurity policies and capabilities result in unintended stockpiles of pathogens and toxins of international security concern. Learning from past outbreaks and pandemics and applying biosecurity measures in conjunction with public health activities is imperative to preventing, detecting, and rapidly responding to emerging and re-emerging health threats.

2.1 The 1918 Influenza Pandemic

The 1918 influenza pandemic is considered the most severe in history, infecting more than one-third of the world's population and killing an estimated 50 million people worldwide. Troops' movement exacerbated the spread of influenza between the spring of 1918 and the summer of 1919 during World War One and advancements in transportation, including ships and railways [36]. Although medical officers of the army isolated soldiers with signs or symptoms, the disease was incredibly contagious, spread rapidly, and infected people in almost every country. Crowd mitigation and quarantines were incredibly difficult to control during the pandemic, given the war constraints, particularly in the training camps and in the battlefields [8].

When the severity of the second wave of influenza became apparent, many countries imposed strict quarantine measures on all incoming ships to prevent the spread of influenza [18]. Health officials in major cities tried to implement multiple strategies to contain the disease, including closing down schools, churches, and theaters, and suspending any large gatherings. People were encouraged to practice respiratory hygiene and social distance, but many of these measures were ineffective since they were implemented too late and often in an uncoordinated way. Other interventions aimed at controlling the spread of the disease, such as travel restrictions and border controls, were often impractical and inefficient, especially in war-torn areas [40]. These situations support the conclusion that biosecurity measures require precoordinated planning during epidemic and pandemic events.

2.2 Ebola West Africa

The 2014–2016 Ebola outbreak in West Africa was the largest and most complex Ebola outbreak since the virus was first discovered in 1976. There were more cases and deaths in this outbreak than all others combined. Weak surveillance systems and poor public health infrastructure contributed to the difficulty surrounding the containment of this outbreak, and it quickly spread to Guinea's bordering countries, Liberia and Sierra Leone. During the epidemic, the Ebola outbreak spread to even more countries worldwide, including countries in Africa, Europe, and the United States. An additional 36 cases and 15 deaths occurred when the outbreak spread outside of these three countries [9].

The Ebola epidemic uncovered significant weaknesses in the global capability for addressing biologic threats. The epidemic highlighted the critical need to build global capacity to prevent, detect, and respond to biological threats and prevent future outbreaks from becoming epidemics. This included the need for safe, secure, and robust laboratories; reliable and timely real-time disease surveillance systems; a well-trained workforce; multi-sectoral collaboration; and functional emergency operations centers to coordinate a successful and rapid response. This outbreak brought forth the idea that biosafety capacity was lacking in Africa, and that biorisk management plays a critical role in the effective epidemic response and the protection of healthcare workers and the community. But with the urgency of the situation, biosecurity was under-addressed.

Throughout the epidemic, many clinical samples were collected and sent to a rapidly established laboratory designed to help with the sudden volume of testing (e.g., mobile laboratories). In addition to the support from donor countries to meet the demands of tests, there was an influx of academic and commercial research organizations conducting trials—all with access to clinical samples. Samples were either destroyed, exported out of an affected country through an official government agreement, exported out of an affected country without an agreement, or remained stored in country. Those samples that continued to be stored in the region were often stored in facilities without biosecurity [1]. To handle Ebola safely and securely requires a Biosafety Level 4 (BSL-4) laboratory; however, there are only two BSL-4 facilities in Africa – Gabon and South Africa [26]. Since the Ebola outbreak ended, there is a multitude of Ebola samples that are either unaccounted for or are stored in facilities that do not have an appropriate level of biosecurity. Like laboratory biosecurity, this epidemic demonstrates the importance of sample and inventory management as a means to control access to unwanted biological materials.

2.3   Ebola DRC

The Democratic Republic of the Congo (DRC) has been grappling with the world's second-largest Ebola epidemic in history since 2018. The realities of fighting an epidemic in a conflict zone make efforts to stop the disease's spread nearly impossible. Violence and political unrest in the affected areas have further restricted the community's access to health care and have undoubtedly caused delays in controlling the outbreak. Security constraints have hindered the Ebola response, posing challenges in identifying new cases, trace contacts, and conducting vital community outreach activities. Some health centers have been damaged or temporarily closed, increasing the risk posed by unsecured samples [2].

Entry points and sanitary control points have been set up in the provinces of North Kivu and Ituri to protect the country's major cities and prevent the spread of the epidemic in neighboring countries. Only four confirmed cases of Ebola have occurred outside of the DRC in Uganda, although there have been no transmission or secondary cases in Uganda. In September 2019, the Commission for Prevention and Biosecurity of the Ministry of Health of the DRC launched a guideline and training package on infection prevention and control targeting nurses, doctors, and other healthcare workers [29]. This effort helped to strengthen health practitioners' proficiency in preventing the spread of Ebola virus disease in health facilities, however, raising awareness among communities about Ebola containment measures has been one of the main challenges of the DRC's Ebola outbreak response. People can be hesitant to accept unfamiliar infection prevention and control practices, such as safe burials or decontamination activities, especially where rumors and information are widespread. Overcoming cultural barries is a significant challenge, underscoring the importance of clear risk communication.

3 Biosecurity Implications of the COVID-19 Pandemic

There is little doubt that the impacts of COVID-19 pandemic will be far-reaching for years, perhaps manifesting a paradigm shift in significant aspects of daily life. This chapter was composed approximately one-year following the initial identification of the SARS-CoV-2 virus outside of China, and yet much has changed in that relatively short period of time. Time will ultimately tell the lasting effects—on our economy, social lives, how we work and travel, changes to healthcare programs and infrastructure, health insurance, and even national security—but it is likely that the post-pandemic world will appear a little different than the world before COVID-19. Here we discuss the applications of biosecurity as a means to augment preparedness and highlight some potential changes in our world.

3.1 Confluence of Infection Control and Biorisk Management

Infection prevention and control is a set of practices and principles aimed at preventing or reducing the spread of infectious agents in healthcare settings. The primary elements that comprise infection control as a discipline and professional field include disinfection and sterilization, environmental infection control, hand hygiene, isolation measures, occupational exposure management, and even reaches into medical devices, common surgical procedures, and organ transplants. The influx of patients, as a result of SARS-CoV-2 infection, has resulted in a major strain on healthcare institutions and entire systems in the U.S. and other countries [13]. Pandemics have the ability to threaten the lives and wellbeing of patients and health-care workers alike. As hospitals work to accommodate a growing population of infected patients, the risk of exposure to the healthcare workforce increases. Further, most hospitals and clinics are not designed with the capability to effectively isolate patients due to their exposure risk potential. The result is a multifaceted issue of capacity, infection control, quality healthcare access, and systemic risk to health-care enterprises. Where infection prevention and control may have appeared some-what programmatic before, the COVID-19 pandemic has thrust this discipline into a significant, understaffed field of practice.

Whereas infection control and prevention is a discipline central to healthcare, biosafety is a set of practices and principles aimed at reducing the potential risk of exposure to biological agents in laboratory settings [23]. Underpinning the field of biosafety are the concepts of containment and risk assessment concept. These concepts and others are detailed in the recently released sixth edition of *Biosafety in Microbiological and Biomedical Laboratories*, the foundation guidelines published and maintained jointly by the U.S. Centers for Disease Control and Prevention and the National Institutes of Health [23]. Of particular value is the biological risk assessment, or biorisk assessment process, that begins with hazard identification and proceeds through appropriate containment measures, levels, and risk communication. Despite availability of true statistics, there are tremendous data to support that wide implementation of biosafety in laboratory environments has dramatically reduced the occurrence of laboratory acquired infections [28, 43].

The two fields of practice share several similarities and overlapping goals. One of several commonalities between infection prevention and control is the concept of containment. In a healthcare setting this usually occurs by isolating the patient that may serve as a source of contagion that could pass to other patients or healthcare workers. In the laboratory, equipment (e.g., biosafety cabinet), engineering controls (e.g. negative pressure laboratory rooms), and personal protective equipment (respiratory protection) serve to protect the laboratory workers from unnecessary exposure to infectious agents. The distinction, of course, is that in the healthcare setting the biological agent is in fact the patient, whereas in the laboratory is may be a sample or specimen of infectious biological materials.

Biocontainment laboratories are specially engineered facilities that provide for controls to reduce the risk of exposure. They range from biosafety level 1 to biosafety level 4 (BSL-1, BSL-2, BSL-3, and BSL-4), where the enhancement of engineering controls increases based on the risk assessment of biological materials and procedures. A hallmark of high-containment biosafety laboratories (e.g., BSL-3, BSL-4) where negative inward airflow, which is often exhausted through HEPA filters, prevents the release of potentially contaminated laboratory air from passing into proximal or adjacent spaces, thereby protecting outsiders. The analogy in the healthcare environment is the patient isolation unit, suite, or room. The rooms have adopted the concept of negative inward airflow to reduce the risk of spreading infections to other areas. Interestingly, the principles of biosafety have had dramatic influences on how laboratories are built, and the concepts of infection prevention will continue to shape how healthcare settings are built. A potential lasting outcome of the COVID-19 pandemic is an increase in the number and rigor of healthcare settings capable of handling and isolating infectious patients.

The most basic of the shared competencies in the two respective disciplines is likely the concept and application of risk assessment and management. At its most fundamental form, the risk assessment and management framework is geared toward identifying those risks to either patients, workers, or public manifested by unnecessary exposure to biological agents. A brief cross-section of shared competencies also reveals exposure control, sterile techniques, biohazard sharps control, waste management, bloodborne pathogen awareness, and basic hand hygiene [12].

Despite such overlap in expertise domains, the inclusion of biosecurity into infection prevention and control is lacking. First, infection prevention and control has areas of focus not associated with biosafety or biorisk management, such as surveillance, epidemiology, and patient care. These particular domains are not often associated with a culture or mindset of security. But biosecurity is likely to encroach into infection prevention as a result of changes in the development and delivery of therapeutics and the significant influence of pandemics. To the former, recent years have seen tremendous promise toward the treatment of prevailing health conditions and disease by way of gene therapy, cell therapies, and other novel biotherapeutics. Not only does this impart biosafety into clinical settings, but it is also likely to bring elements of biosecurity as well. These therapies are often patient-specific (e.g., developed for a single patient) and are quite expensive. In essence, they are valuable biological assets with a sensitive supply chain.

It is no surprise to practitioners of public health and health security that biorisk management (biosafety and biosecurity) is a contribution of overall health security. Biorisk management is a skillset required by all laboratories at the front lines of epidemic and pandemic threats. Yet, the GHSA position on biosecurity is largely relegated to the laboratory levels. Future efforts will be required that take the precepts of biosecurity application and expand them outside of the laboratory, not wholly unlike the widely adopted methods of agricultural biosecurity. As discussed, perhaps the COVID-19 pandemic will institute cross-collaboration amongst related disciplines such as infection control, infection prevention, industrial hygiene, and continued overlaps with biosafety. To date, many of these disciplines have been narrowly applied and operate independent of one another.

3.2 Health Security Will Become More Common

The term "health security" implies that the health of an individual is a component of the overall sense of security of the individual. This is applicable to populations as well. As a collection of societies, and a species, we recognize that threats to our health can jeopardize our security, and that threats to our security can compromise our health. Further, the health of our populations is worthy of imparting security measures- health has implicit value and is therefore worth protecting. The COVID-19 pandemic is a current example that our health and our security are intertwined.

From the government perspective, health is integral to national security. The health of its citizens is directly associated with the health of the economy: a sick population can't work and contribute to economic viability or growth. Failure to contain outbreaks of infectious disease can result in collapse of commerce, import, export, tourism, agriculture, and travel, all which are vital to economic stability and growth [15, 38]. Failing economies are a prerequisite for civil instability and increased vulnerabilities from outside influences or nonstate actors. Extrapolating this premise to a global level results in the foundation of the GHSA (GHSA, Chap. "Biodefense Promotes Biosecurity Through Threat Reduction Programs and Global Health Security") [6].

From the individual level, health is another indicator of personal security. An ill individual may not be able to conduct their job, leading to personal financial losses. The COVID-19 pandemic has made clear this connection. A number of autonomous zones have manifested in major U.S. cities during the course of the pandemic, and clashes between citizens and law enforcement have occurred, in part due to the issue of eviction [21]. If citizens are too ill to work and find themselves unemployed, or unemployment rises suddenly and dramatically, civil unrest is somewhat predictable. The COVID-19 pandemic has resulted in new unemployment levels and raised the level of poverty, thereby promoting elements of civil instability. For these reasons, the connection between health and security at an individual or family level are clear and fragile.

There remains no federal law in the U.S. that ensures workers can have or earn paid sick days, and access to paid sick days from employers is slanted toward middle- to high-earning jobs [17]. Many economic models, however, indicate paid sick days actually promote stronger economies [22]. Yet, the U.S. and other countries face systemic issues of employees reporting to work while ill to prevent losing their jobs. It is possible that one lasting effect of the COVID-19 pandemic is a reduction in employees coming to work when they are ill. This is likely due to the stigma attached with being ill, or being perceived to be ill, as a result of the pandemic. Social stigma with illness is well documented as a result of the decades-long HIV/AIDS epidemic [20]. Perhaps the COVID-19 pandemic will help reduce the pressures, and resulting economic losses, of sick employees going to work and infecting additional workers. In this event, governments must be prepared to reconsider the availability and equity of access to paid sick leave as a means to promote individual and population health security.

The COVID-19 pandemic will certainly not be the last pandemic to demonstrate that the health of our species is integral to our individual and collective sense of security. In a way, the pandemic has brought forth a reality that underpins the combined efforts exemplified in the GHSA. The premise is simple: a single country cannot expect to prepare itself from the influx of infectious disease, regardless of its country of origin. Instead, arming each country with capacities and capabilities to prevent, diagnose, and respond effectively is similar to ensuring that each link in a chain is sufficiently strong. The COVID-19 pandemic has starkly revealed the tenuous connection between health, employment, financial stability, and therefore security. Although speculative it is possible that the COVID-19 pandemic will also bring closer the concepts of health and security at an individual level.

3.3 Bioterrorism May Increase

The COVID-19 pandemic has imparted catastrophic losses economically, socially, and in the toll of human life. Notably, the SARS-CoV-2 virus, while highly transmissible, it not highly virulent. This indicates that mortality via virulence is not necessary for an infectious disease or biological agents to have devastating effects on human health, societal structures, and the economy. The COVID-19 pandemic has issued a masterclass in how a biological threat can test the interconnectedness of our world and species. In short, the world has truly taken notice and seen the evidence of the damages.

The effects of the COVID-19 pandemic resemble the aspirations of many would-be malicious actors and groups that may consider the use of biological weapons. In short, the COVID-19 pandemic has provided ample proof of the terror and consequences that infectious agents can have. Is it possible that existing and developing terrorist organizations, that may once have been undecided on the pursuit of biological weapons, are now reconsidering such a weapon?

The COVID-19 pandemic has also demonstrated that many of the population are already sensitized to the false notion that SARS-CoV-2 was engineered, released intentionally, or escaped from a laboratory [27, 33, 34]. This is despite overwhelming consensus between scientists and intelligence agencies that these beliefs are categorically false [19]. It is likely that, like many false or conspiratorial theories, these beliefs are rooted in fear. While the conspiracy theories of the SARS-CoV-2 virus are false, the means and opportunity to engineer other biological agents to manifest fear or panic are real. Based on these reasons it is likely that the threshold to approach bioterrorism as a means to impart terror is likely lower as a result of the COVID-19 pandemic. This is currently the subject of great discussion with calls for international cooperation to assess and mitigate the potential for the increase threat of a bioterrorism attack. For example, in the middle of 2020 the Council of Europe, a human rights organization with 47 member states, convened to discuss and chart a path forward the potential for an increase in bioterrorism following the COVID-19 pandemic. This may be one of the first organized meetings to discuss this emerging topic, but it will be followed by many more. Bioterrorists have now clearly observed that "it will work." If this is the case, how then can biosecurity be applied to prevent, mitigate, respond, and recover to a true biological attack?

Answering this question relies on our ability to follow the central dogma of the biological threat and risk assessment detailed in Chap. "The Biothreat Assessment as a Foundation for Biosecurity," but at a much larger level than a single laboratory or enterprise. This process begins with prioritizing the assets. At a national level this is a massive undertaking, particularly when one considers the breadth of assets in all things biological: food, medicines, research, innovation, infrastructure, economy, and others. We must be able to assess the integral components of the entire landscape of biology and biomedicine that contributes to the health, safety, security, and prosperity of the world's populations. As we have described, the definitions of biosecurity vary based on the lens through which they are often narrowly applied. Laboratory biosecurity, for example, is aimed at protecting valuable or dangerous biological assets in a laboratory environment. Further assessment is required to determine if those same principles can be adapted to a different environment, for example a site where clinical trials for vaccines are being conducted. Given the high-profile, financial investment, and exponential downstream losses it is reasonable to assume that the multitude of clinical trials being conducted in response to an emerging infectious disease would present as a suitable target of bioterrorism. Today, most security considerations and expenditures associated with clinical trials are limited to patient data and information security. Similarly, the supply chain of biological products, from food to vaccines, offer a variety of vulnerabilities that are difficult to secure [7].

The second major category is working to characterize those entities, groups, or individuals that with the ability to potentially use a bioweapon. This is effectively the efforts required to identify and characterize threats. Every day, scores of intelligence agencies around the world search, track, identify, and collect information on threats of all types. This includes transnational terrorist organizations like Al-Qaeda to individual cybercriminals making use of dark web technologies. For example, Interpol maintains a Counter-Terrorism Directorate that serves as a global hub for intelligence on transnational terrorists with numerous partner countries. In the U.S., the Federal Bureau of Investigation (FBI) and the Central Intelligence Agency have counter-terrorism units dedicated to monitoring terrorist activities and networks. Despite this, there is lacking data to suggest that the use of behavioral threat assessment is used to evaluate profile of terrorists or criminals that may be drawn to bioweapons over more common weapons or weapons of mass destruction. As Chap. "The Biothreat Assessment as a Foundation for Biosecurity" details, the FBI has recently adopted threat assessment as a field of practice to support the identification, assessment, and management of targeted attacks. Yet the psychological and behavioral parameters of the "bioterrorist" remain largely undifferentiated from other types of terrorists. Perhaps additional study in this field will support proactive threat assessment and threat characterization for law enforcement agencies around the world with respect to bioterrorism.

Identification and characterization of vulnerabilities across the spectrum of economy of biological assets is a daunting task, but to date many efforts are already focused on this. A prime example in the U.S. is the Bipartisan Commission on Biodefense which was established to identify "systemic weaknesses in the national biodefense posture and recommend ... steps the government can take to mitigate [vulnerabilities]" [4]. This Commission is privately funded and serves to assist and guide legislators, whereas the U.S. government has departments and agencies focused on assessing vulnerabilities as well. These include the U.S. Department of Homeland Security's Cybersecurity and Infrastructure Security Agency, the U.S. Department of Agriculture's Food Safety and Inspection Service, and the U.S. Department of Defense Threat Reduction Agency. These Departments often work collaboratively to share information on overlapping vulnerabilities. Repairing vulnerabilities is an exercise in building resiliency, which is a primary theme of the GHSA (Chap. "Biodefense Promotes Biosecurity Through Threat Reduction Programs and Global Health

Security"). Here, vulnerabilities are identified and managed across a spectrum of action packages aimed at augmenting capacities and capabilities to mitigate natural and manmade biological events. These are but a few examples of national or international programs with the intent of managing vulnerabilities that can be exploited by bioterrorists.

It is the intersection of threats and vulnerabilities that risks result. Despite the best efforts, and the vast lessons learned from the COVID-19 pandemic, it is not possible to deter all threats and correct all vulnerabilities. The terrorism landscape is simply too dynamic. Therefore, great emphasis should also be placed on response and recovery efforts. The COVID-19 pandemic has demonstrated that national and international level plans are required. In the event of a bioterrorist attack that imparts similar scope and scale of catastrophe, it is arguable that our countries will require a higher level of preparation. This may take the form of new or augmented national-level plans for response, greater collaboration and information sharing across Departments, Ministries, and Agencies, greater investment into counter-terrorism services, assessing infrastructure and supply chains, and reevaluating the mission of respective defense agencies and programs to include more emphasis on bioterrorism. Regardless, time will reveal if the number and diversity of bioterrorist activities occur after the COVID-19 pandemic comes to an end.

3.4 Cyber Criminals Will Have New Targets

Cybersecurity threats are some of the fastest growing acts of crime in our world today with estimates placing economic losses due to cybercrimes in excess of $10 trillion annually by 2025 [24]. The year 2020 has seen an exponential rise in ransomware cybercrimes with devastating effects on all types of organizations [37]. The COVID-19 pandemic has unfortunately presented a most lucrative digital environment for cybercrimes for several reasons.

One, the onset of lockdown conditions forced many to switch to their homes as their primary workplace. For most office workers in the U.S. and other developed countries this was relatively a simple switch in daily routines. However, most homes have far weaker IT security platforms, programs, and controls than do employers. This provides for increased vulnerabilities compared to professional environments. Employers are often less prepared to defend and triage the respective networks of a remote workforce. Secondly, the public is highly sensitized to any and all news about the pandemic, which has dominated newsfeeds since early 2020. This has led to a growing number of phishing schemes and ransomware lures that have caused people to click false internet links and website. In this respect, the COVID-19 pandemic has created new avenues for cyber criminals to exploit [14].

An emerging domain that brings together the tenets of cybersecurity and the need to protect the economy of the biological industries ("bioeconomy") is termed cyberbiosecurity [25, 31, 32]. Disruption of the bioeconomy can have major impacts on agriculture, patient care, research and development, therapeutics development, vaccines, and others. Virtually all fields of biology are converging with some aspects of information technology which opens the bioeconomy to cybersecurity threats. The COVID-19 pandemic has demonstrated strategic and escalated efforts with respect to patient care and the development of SARS-CoV-2 vaccines, and cybercrimes against these two areas have risen [14, 35].

Hospitals are a growing target of cybercriminals during the COVID-19 pandemic thus far and patient medical records are of growing concern. Ransomware attacks on healthcare enterprises climbed 350% toward the end of 2019 versus the same time in 2018, and this escalation is anticipated to grow [5]. 2020 has seen in excess of 5.6 million patient records that have been breached and compromised as a result of external cybersecurity breaches [11]. It is not unusual that during the surge associated with an epidemic or pandemic event creates pressure healthcare providers to accommodate a mass number of patients, with a focus on patient care. The result is that cybersecurity concerns are not paramount, in addition to reports that healthcare providers typically spend two- to three-times less on cybersecurity than do other organizations that handle sensitive personal data [5]. But the attacks go beyond simply hacking hospitals, obtaining patient data, and selling it on the dark web. Düsseldorf University Hospital fell victim to hackers that shut down all computer systems. This resulted in an incoming, critical patient being delayed and diverted to another hospital. During this transfer the patient died, and although the investigation is not concluded, this may mark the first time that a cyberattack resulted in a death: this investigation is being labeled as a homicide by German officials [39].

Cybercriminals are also aptly aware of the significant value surrounding the intellectual property being developed by vaccine producers as they work to research and manufacture SARS-CoV-2 vaccines. The global pressure of the pandemic has thrust vaccine development and production into a pace unseen with prior vaccines, and this has exposed potential vulnerabilities. In December of 2020, hackers breached information systems at the European Medicines Agency, the vaccine regulatory body for many European nations, and accessed data related to the oversight and regulatory status of the SARS-CoV-2 vaccine [16]. Earlier in July of 2020, the U.K.'s National Cyber Security Centre reported a hacking group with reported ties to Russian intelligence services has launched several attacks targeting academic research centers and pharmaceutical companies, despite Russia's denial of knowledge of the attacks [42]. Shortly after this incident, the U.S. Department of Justice filed an indictment against two Chinese nationals accused of hacking multiple companies, governments, and political forums, including Novavax and Moderna, two companies that were working toward development of SARS-CoV-2 vaccines [10]. Other known attacks have targeted Johnson & Johnson, AstraZeneca, and other leading SARS-CoV-2 vaccine developers [16]. Although relatively early, the true damage caused by cybercriminals with respect to SARS-CoV-2 research and vaccine development will be calculated as time passes.

Cybercrime is not limited to hacking COVID-19 information related to patients and vaccine production. There are reported attacks designed to also interrupt the supply chain as vaccines are prepared for delivery and distribution [3]. Two of the SARS-CoV-2 vaccines that are slated for emergency regulatory approval in the U.K.,

Canada, and the U.S. require sustained subzero temperatures during the shipping and storage processes adding considerable logistical challenges for widespread distribution. Cybersecurity researchers at IBM started to gather intelligence as early as September 2020, well before any physical distribution of vaccine shipments, indicating that hackers were working ahead of the distribution campaigns to develop their attacks [3]. To date it is unclear if any of the attacks were successful at compromising organizations involved in the logistical deployment of the vaccines.

Unlike many vaccines that developed in the past, several of the SARS-CoV-2 vaccines are mRNA vaccines reliant upon the integrity of the genetic sequences to produce the correct immune-responsive molecules in patients. A recent trend in cybercrime is the use of traditional hacking methods to breach a network or computer, using a computer virus that alters the target genetic sequence into something that results in a harmful or pathogenic product [30]. This has been identified as a major potential threat to the bioeconomy as the synthetic biology industry manufactures and sells billions of sequences to customers annually. This type of threat is also occurring ahead of any single, central database on which genetic sequences can be screened for pathogenic elements. As the field of vaccine development continues to make use of genetic technologies it is clear to understand the potential for devastating effects.

4 Summary

To date, biosecurity has largely been an institutional practice with historic emphasis on laboratories and agriculture. Here, we have examined a few of the potential lasting effects of the COVID-19 pandemic and offered mechanisms in which the principles of biosecurity can be extended beyond the laboratory or farm to augment other sectors that may play a role in detection, deterrence, response, and recovery of the next natural or intentional biological event.

The authors note that greater collaboration and convergence of related disciplines, such as biorisk management and infection prevention, must occur in an orchestrated, deliberate manner to capitalize on respective competencies. Just as the SARS-CoV-2 virus ignores boundaries and silos, so too must professionals with a stake in public health. The GHSA offers an opportunity for great capacity building, particularly in those nations that are under-resourced and lack current capabilities to contain an outbreak (or prevent a bioterrorist event) resulting in a spread to other countries or around the world. The ideals of health security will become more individualized and personal. Perhaps this offers an opportunity to weave science- and risk-based approaches into how populations live and prosper every day.

The maturity of the bioeconomy suggests that there is more at stake than ever before to prevent unwanted attacks that jeopardize our food, medicines, and ultimately our societies. If the use of an epidemic agent is an unorthodox weapon, we too should be looking to protect unorthodox assets that may fall victim to a bioterrorist attack. The definition of biosecurity that focuses only on the laboratory environment is inadequate to secure the billions of dollars and billions of lives dependent on the fruits of the biological and biomedical sciences. It stands to reason that valuable biological assets extend beyond a vaccine or DNA sequence into nontraditional areas. These include supply chains of critical therapeutics, clinical trial sites where patients are protected and vulnerable, protecting the integrity of the burgeoning synthetic biology enterprise, rethinking how security is applied to healthcare entities- all of which have existing, documented vulnerabilities subject to bioterrorist exploitation. Biosecurity and biodefense practitioners are required now, in the post-pandemic landscape, to adjust perspectives adequately to objectively understand the biosecurity applications in novel areas. Although we will not know the true toll of the COVID-19 pandemic for years to come, the authors feel adequate lessons and wisdom are currently being amassed to begin to prepare for the next pandemic.

References

1. Abayomi A, Katz R, Spence S, Conton B, Gevao SM (2016) Managing dangerous pathogens: challenges in the wake of the recent West African Ebola outbreak. Glob Secur: Health Sci Policy 1(1):51–57. https://doi.org/10.1080/23779497.2016.1228431
2. Aruna A, Mbala P, Minikulu L, Mukadi D, Bulemfu D, Edidi F, et al (2019) Ebola virus disease outbreakdemocratic republic of the Congo, August 2018–November 2019. MMWR. Morb Mortal Wkly Rep 68(50):1162–1165. https://doi.org/10.15585/mmwr.mm6850a3
3. Barrett, B (2020) Hackers are targeting the COVID-19 vaccine 'cold chain.' Wired. https://www.wired.com/story/hackers-targeting-covid-19-vaccine-cold-chain/
4. Bipartisan Commission on Biodefense (2015) A national blueprint for biodefense: leadership and major reform needed to optimize efforts. Bipartisan Commission on Biodefense c/o Hudson Institute, Washington, D.C. https://biodefensecommission.org/wp-content/uploads/2015/10/NationalBluePrintNov2018-03.pdf
5. Bobo, B (2020) COVID-19 and health care cybersecurity: how to protect practices and patient data. Med Econ. https://www.medicaleconomics.com/view/covid-19-and-cybersecurity-protect-practices-and-patient-data
6. Bunnell RE, Ahmed Z, Ramsden M, Rapposelli K, Walter-Garcia M, Sharmin E, Knight N (2019) Global health security: protecting the United States in an interconnected world. Public Health Rep 134(1):3–10. https://doi.org/10.1177/0033354918808313
7. Burnette R (2013) Biosecurity: understanding, assessing, and preventing the threat. Wiley, Hoboken, NJ
8. Byerly CR (2010) The U.S. military and the influenza pandemic of 1918–1919. Public Health Rep (Washington, D.C., 1974), 125(Suppl 3):82–91
9. Coltart CEM, et al (2017) The Ebola outbreak, 2013–2016: old lessons for new epidemics. Philos Trans R Soc London Ser B Biol Sci 372(1721):20160297. https://doi.org/10.1098/rstb.2016.0297
10. Cox K (2020). DOJ: Chinese hackers stole "hundreds of millions of dollars" of secrets. ArsTechnica. https://arstechnica.com/tech-policy/2020/07/doj-accuses-chinese-hackers-of-trying-to-steal-covid-19-research-data/
11. Davis J (2020). COVID-19 Impact on ransomware, threats, healthcare cybesecurity. cybersecurity news. https://healthitsecurity.com/news/covid-19-impact-on-ransomware-threats-health care-cybersecurity
12. Emery R, Patlovich S, King K, Lowe J, Rios J (2016) Comparing the established competency categories of the biosafety and infection prevention professions: a possible roadmap

for addressing professional development training need for a new era. Appl Biosafe. June; 21(2):79–83. https://doi.org/10.1177/1535676016651250

13. Evans, M. (2020, November 11). *Record COVID-19 Hospitalizations Strain System Again*. The Wall Street Journal. Retrieved from: https://www.wsj.com/articles/covid-19-surge-strains-hos pitals-once-again-11605100312

14. Ferbrache D (2020) The rise of ransomware during COVID-19: how to adapt to the new threat environment. KPMG. https://home.kpmg/xx/en/home/insights/2020/05/rise-of-ransom ware-during-covid-19.html

15. Fitzmaurice AG, Mahar M, Moriarty LF (2017) Contributions of the US Centers for Disease Control and Prevention in implementing the Global Health Security Agenda in 17 partner countries. Emerg Infect Dis 23(13)

16. Goodin D (2020). COVID-19 vaccine data has been unlawfully accessed in hack of EU regulator. ArsTechnica. https://arstechnica.com/information-technology/2020/12/hackers-unl awfully-access-data-related-to-promising-covid-19-vaccines/

17. Gould E, Schieder J (2017) Work sick of lose pay? The high cost of being sick when you don't get paid sick days. Economic policy institute, Washington, D.C. https://epi.org/130245

18. Johnson N. (2006) Britain and the 1918–1919 influenza pandemic: a dark epilogue. Taylor & Francis Ltd, Abingdon

19. Lewis T (2020) Eight persistent COVID-19 Myths and why people believe them. Sci Am. https://www.scientificamerican.com/article/eight-persistent-covid-19-myths-and-why-peo ple-believe-them/

20. Kontomanolis EN, Michalopoulos S, Gkasdaris G, Fasoulakis Z (2017) The social stigma of HIV-AIDS: society's role. HIV/AIDS (Auckland, NZ) 9:111–118. https://doi.org/10.2147/ HIV.S129992

21. Marcus E (2020) In the autonomous zones. The New York Times, New York. https://www.nyt imes.com/2020/07/01/style/autonomous-zone-anarchist-community.html

22. Marotta J, Greene S (2019) Paid sick days: what does the research tell us about the effectiveness of local action? Urban Institute, Washington, D.C. https://www.urban.org/sites/default/files/ publication/99648/paid_sick_days._what_does_the_research_tell_us_about_the_effective ness_of_local_action_0.pdf

23. Meechan P, Potts J (2020) Biosafety in microbiological and biomedical laboratories (6th ed.). U.S. Dept. of Health and Human Services, Public Health Service, Centers for Disease Control and Prevention, National Institutes of Health, Washington, D.C.

24. Morgan S (2020) Cybercrime to cost the world $10.5 Trillion annually by 2025. Cybercrime Mag. https://cybersecurityventures.com/hackerpocalypse-cybercrime-report-2016/

25. Murch RS, So WK, Buchholz WG, Raman S, Peccoud J (2018) Cyberbiosecurity: an emerging new discipline to help safeguard the bioeconomy. Front Bioeng Biotechnol 6:39. https://doi. org/10.3389/fbioe.2018.00039

26. National Academy of Sciences (2011) Biosecurity challenges of the global expansion of high containment biological laboratories. National Academies Press (US), National Research Council Committee on Anticipating Biosecurity Challenges of the Global Expansion of High-Containment Biological Laboratories. 8, Requirements for and challenges associated with BSL-4 Labs (plenary session), Washington, DC. https://www.ncbi.nlm.nih.gov/books/NBK 196156/

27. Nie JB (2020) In the shadow of biological warfare: conspiracy theories on the origins of COVID-19 and enhancing global governance of biosafety as a matter of urgency. J Bioeth Inq 17(4):567–574. https://doi.org/10.1007/s11673-020-10025-8

28. Peng H, Bilal M, Iqbal HMN (2018) Improved biosafety and biosecurity measures and/or strategies to tackle laboratory-acquired infections and related risks. Int J Environ Res Public Health 15:2697

29. ProMED (2019). Ebola virus spread continues in DRC–CEPI. https://cepi.net/news_cepi/dis ease-outbreak-update-ebola-virus-spread-continues-in-drc/

30. Puzis R, Farbiash D, Brodt O et al (2020) Increased cyber-biosecurity for DNA synthesis. Nat Biotechnol 38:1379–1381. https://doi.org/10.1038/s41587-020-00761-y

31. Richardson L, Connell N, Lewis S, Pauwels E, Murch R (2019) Cyberbiosecurity: a call for cooperation in a new threat landscape. Front Bioeng Biotechnol 7:99. https://doi.org/10.3389/fbioe.2019.00099

32. Richardson L, Lewis S, Burnette R (2019) Building Capacity for Cyberbiosecurity Training. Front. Bioeng. Biotechnol. 7:112. https://doi.org/10.3389/fbioe.2019.00112

33. Schaeffer K (2020a) A look at the Americans who believe there is some truth to the conspiracy theory that COVID-19 was planned. Pew Research Center. https://www.pewresearch.org/fact-tank/2020/07/24/a-look-at-the-americans-who-believe-there-is-some-truth-to-the-conspiracy-theory-that-covid-19-was-planned/

34. Schaeffer K. (2020b) Nearly three-in-ten Americans believe COVID-19 was made in a lab. Pew Research Center. Retrieved from: https://www.pewresearch.org/fact-tank/2020/04/08/nearly-three-in-ten-americans-believe-covid-19-was-made-in-a-lab/

35. Seals T (2020) Healthcare 2021: cyberattacks to center on COVID-19 spying, patient data. Threat Post. https://threatpost.com/healthcare-2021-cyberattacks-covid-19-patient-data/161776/

36. Short KR, Kedzierska K, van de Sandt CE (2018) Back to the future: lessons learned from the 1918 influenza pandemic. Front Cell Infect Microbiol 8:343https://doi.org/10.3389/fcimb.2018.00343

37. Singleton C, Kiefer C, Villadsen O (2020). Ransomware 2020: attack trends affecting organizations worldwide. Secur Intell. https://securityintelligence.com/posts/ransomware-2020-attack-trends-new-techniques-affecting-organizations-worldwide/

38. Tappero JW, Cassell CH, Bunnell RE. (2017) US Centers for Disease Control and Prevention and its partners' contributions to global health security. Emerg Infect Dis. 23(13)

39. Tidy J (2020). Police launch homicide inquiry after German hospital hack. BBC News. https://www.bbc.com/news/technology-54204356#:~:text=German%20police%20have%20launched%20a,transfer%20her%20to%20another%20hospital

40. Tognotti E (2013) Lessons from the history of quarantine, from plague to influenza A. Emerg Infect Dis 19(2):254–259. https://doi.org/10.3201/eid1902.120312

41. World Health Organization (2020) Ebola virus disease: democratic republic of Congo. Extern Situat Rep 98. https://www.who.int/emergencies/diseases/ebola/drc-2019/situation-reports

42. Warrell H, Cookson C, Foy H (2020) Russia-linked hackers accused of targeting COVID-19 vaccine developers. ArsTechnica. https://arstechnica.com/information-technology/2020/07/russia-linked-hackers-accused-of-targeting-covid-19-vaccine-developers/

43. Wurtz N, Papa A, Hukic M, Di Caro A, Leparc-Goffart I, Leroy E, et al (2016) Survey of laboratory-acquired infections around the world in biosafety level 3 and 4 laboratories. Eur J Clin Microbiol & Infect Dis 35(8):1247–1258

The Changing Face of Biological Research and the Growing Role of Biosecurity

Nicolas Dunaway and Kavita M. Berger

Abstract The landscape of biological research is an ever evolving one. From the discovery of bacteriophages in 1915 to the reprogramming and assembly of organisms in 1972, the 20th century was one of the great breakthroughs that transformed the way biology is studied and applied to real-world problems. Advances in biology and biotechnology have provided new capabilities for addressing social needs in health, agriculture, energy, environment, and defense. The pace of scientific and technological advancement is accelerating rapidly with the inclusion of new scientific disciplines (e.g., material and information sciences), non-traditional practitioners (e.g., community laboratories and engineers), and new funding mechanisms (e.g., crossover venture capital and crowdsourcing). Since the start of the twenty-first century, several discoveries and technological breakthroughs far outpace the wildest of imaginations of the twentieth century. These technology developments offer great hope for biodefense and global health security, but they also have the potential for increasing biosecurity risks. In this chapter, we explore the biotechnology landscape, including scientific communication, advances and applications, practitioners, and funders.

1 Developing Technologies and Biosecurity Considerations

At the beginning of the 21st century, the security community marveled at the capabilities offered by synthetic biology and associated fields. Synthetic biology is a multidisciplinary field that applies an engineering approach to biology where biological materials or systems are designed and synthetically developed for research and/or functional applications. The concept builds on forty years of genetic engineering

N. Dunaway
Merrick & Company, Arlington, Washington, D.C., USA

K. M. Berger (✉)
Gryphon Scientific LLC, Takoma Park, USA
e-mail: kberger@nas.edu

National Academies of Sciences, Engineering, and Medicine, Washington, D.C., USA

© Springer Nature Switzerland AG 2021
R. N. Burnette (ed.), *Applied Biosecurity: Global Health, Biodefense, and Developing Technologies*, Advanced Sciences and Technologies for Security Applications, https://doi.org/10.1007/978-3-030-69464-7_6

improvements, leverages new advances in DNA synthesis and recombinant DNA technologies, and applies design-based problem-solving approaches to biology. This field has enabled amateur scientists to explore new applications of biology, educate the public, and become empowered to conduct scientific studies. These events share a close connection with the explosive growth and decreased cost of computer science, electrical engineering, information technology, and the internet. The ability of citizen and professional scientists to use software to design new organisms has opened the door to the use of biology as a vehicle for a variety of uses, including data storage, chemical production, and the production of synthetic and novel molecules and organisms.

These technological developments also have affected the social, psychological, and cultural considerations of biological research, which are pushing the boundaries of activities that are societally acceptable. Recent studies have described cloning of monkeys, precise editing of live human embryos, and the creation and environmental release of sterile insects. Although these activities were once considered unethical, today, the international drivers for these studies are becoming increasingly pervasive. Within the context of national security, these and other scientific advances challenge current norms for responsible scientific exploration. The biological research is global with scientists engaging in a numerous life science and biotechnology fields, and accessing and contributing to international scientific literature. Furthermore, biotechnology developments, such as synthetic biology and artificial intelligence, have redefined what is possible in biological and biotechnological research.

2 Opportunities and Concern About Open Access Publication

Leveraging advances in science and technology is predicated upon the sharing of information about new discoveries, developments, and applications. A brief description of key aspects of scientific communication that promote the exchange of new ideas and knowledge and peer validation of previous findings is important for appreciating how and why the recent move towards open access publication is changing the biotechnology landscape today.

2.1 Scientific Communication and Peer Review

The fundamental question of how to share and distribute new scientific discoveries has been asked by scientists for centuries. Documenting and sharing scientific discoveries through writing started in the 17th century, which laid the foundation for critical evaluation, validation, and discussion about new findings among the scientific elite [60]. These efforts eventually led to the development of a formal method of scholarly refereeing, which is more commonly known as *peer review*. The basic concepts of peer review are rather simple: the author of the written scholarly work submits, in writing, evidence of scientific exploration to peers recognized as "experts" in the same field. If these experts determine the ideas, research, or conclusions included in the submitted document are valid, important, and adds to the body of scientific knowledge, the document is published formally and distributed for access by other scientists and interested individuals. This process aims to provide an impartial and accurate system for ensuring only rigorous and validated scientific research is published.

Although this method has been used with success since the 17th century, it is not without its weaknesses [5]. The rate of publication in traditional scientific journals worldwide has grown steadily since the end of the 20th century, with estimates that the global scientific output doubles every nine years [73, 74, 133]. Furthermore, the number and diversity of journals have increased dramatically during the past 30 years, in large part to promote the publication of research findings in specific fields of study, sectors, or biotechnologies. For example, journals on biosensors, biophysics, and CRISPR-based genome editing exist today to provide a forum in which scientists interested in these fields can learn about new discoveries, applications, and advances. Although the expansion of scholarly journals increases the amount of information shared, some scientists have suggested that this growth has resulted in a decrease in the quality of the peer review, and consequently, the published science [14, 39, 137]. This concern is exacerbated by the vast amounts of information shared across the internet, where the quality and accuracy of the scientific claims are difficult to assess. Today, scientists are responsible for determining what information has been produced using defensible experimental and analytical approaches, which can be challenging given the large amounts of information available. Scientific journals and various organizations have initiated an effort to improve the reproducibility of studies, results, and claims.

2.2 Accessibility of Scientific Information and Open Access Publication

In general, many scientists pursue biological research to help address a societal need, whether that need is in improved crop quality and disease resistance, preserving biodiversity, or human health. Biotechnologies are developed to address technical challenges in this pursuit of societal benefit. Illustrative examples include 3D bioprinting of organs, big data analytics for pathogen early warning and surveillance, and precise genome editing for improved crop health. The benefits of biological and biotechnological advances rely on broader access to published scientific literature, primarily as a way of engaging various stakeholders of the research, including citizen scientists, health practitioners, venture capitalists, sustainability and development experts, and security experts. The importance of sharing information among researchers and other stakeholders has led research foundations, government entities, and scientific journals to promote open access publication of research. The U.S., Canada, and Europe have established PubMed Central databases of research articles that are accessible to users. The U.S. National Institutes of Health has a data sharing requirement, which requires grantees to submit manuscripts to the U.S. PubMed Central to enable publication access. Similarly, the Bill and Melinda Gates Foundation requires their grantees to make their data and manuscripts accessible to stakeholders. Finally, open access journals (e.g., PLoS eLife, Frontiers) have been created to promote accessibility and sharing of scientific information. For example, Howard Hughes Medical Institute, Welcome Trust, and Max Plank Society created eLife, an online open access journal [37]. Although several journals still require paid access, some including, *Science*, have created versions that are open access (*Science Advances*). The cost of publishing in open access journals often is borne by the authors rather than readers. However, new models that involve temporary access to articles (e.g., renting articles and subscription models) have emerged.

A particularly illustrative example of promoting open access involves the work of Mr. Aaron Swartz, whose talents led to the development of open access platforms that are used by biologists and biotechnologists. Born in 1986, Swartz demonstrated exceptional gifts in software engineering, computer science, and the Internet. At age 14, Schwartz became a member of the working group that developed RSS 1.0, a technology that retrieves and aggregates information from the vast array of news sources on the internet, making vast quantities of information available to anyone [57]. This technology became a staple resource in every sector and discipline, enabling stakeholders to read and access information about their topics or fields of interest. As an entrepreneur, Swartz became involved in the development and creation of numerous technologies and organizations that seek to increase accessibility and sharing of information to the broader society, including the Creative Commons framework. The Creative Commons is a licensing framework that provides individuals with a means to "legally share your knowledge and creativity to build a more equitable, accessible, and innovative world" [29]. Later in life, Schwartz founded Demand Progress, a charitable organization that focuses on affecting political change by promoting internet freedom and open government policies in the U.S. [33]. Swartz went on to develop SecureDrop, a tool for whistleblowers to anonymously share information with journalists [68, 105]. In Swartz's landmark 2008 document, the "Guerilla Open Access Manifesto," he makes a definitive statement about his vision of the open access movement, which is to "ensure that scientists do not sign their copyrights away but instead ensure their work is published on the internet, under terms that allow anyone to access it." Swartz issues a call to action for open sharing of information, "Those with access to these resources—students, librarians, scientists… need not—indeed, morally, you cannot—keep this privilege for yourselves. You have a duty to share it with the world" [128]. In publishing this document on the internet, Swartz was making the profound statement that sharing information is a moral imperative. This desire to share information remains consistent with the concept of open publication of fundamental research, which is protected from classification in the United States by a 1985 Presidential directive (National Security Decision Directive-189). The overlapping issues of open publication of scientific advances and open access are critical considerations in current biosecurity policy discussions, particularly discussions on the dual-use dilemma.

2.3 Bridging Open Access and Peer Review

Efforts to promote open access in science recently have been adapted to biological research, providing opportunities for improving the informal peer review processes. An early example of these is arXiv, which was created in 1991 and is an open access, online archive and distribution service that focuses on physics, mathematics, and computer science articles [4]. This platform was created, in part, to address concerns about scientific advancements leading to national security vulnerabilities [27]. In 2013, this model of pre-publication of research was adapted for the biological sciences through the creation of bioRχiv primarily to enable open access and peer review rather than address potential national security vulnerabilities [24]. Subsequently, the model was used to create medRχiv [24]. These services do not conduct peer-review. Instead, they provide researchers in the life sciences with opportunities for making their findings available to the scientific community immediately. they also provide a method for these researchers to receive feedback on draft manuscripts before they are submitted to peer-reviewed journals. These service have become a place where scientists can comment on and/or critique published research if concerns about quality and accuracy are raised. Therefore, these services are looking to support, rather than replace, scientific journals.

Another platform for promoting the accessibility of scientific information is CiteSeer^X which is a digital library that aims to "improve the dissemination of scientific literature and to provide improvements in functionality, usability, availability, cost, comprehensiveness, efficiency, and timeliness in the access of scientific and scholarly knowledge" [23]. To achieve this goal, the service provides "resources such as algorithms, data, metadata, services, techniques, and software that can be used to promote other digital libraries."

Although these examples support and bolster the effectiveness of the traditional, peer-reviewed journal publication process, new efforts have begun to change the informal peer review process by leveraging social networks and crowdsourcing. Through these means, scientists are encouraged to review, critique, and organize scientific activities more efficiently and accurately than the traditional peer review process. These efforts have the potential to revolutionize the system of scientific communication. However, at least one effort—the comment section of PubMed— has not generated the level of dialogue envisioned, which led the U.S. NIH to end the service. Several challenges are associated with implementing this crowd-based review system into practice: (1) getting the requisite number of people involved to reap the benefits of crowdsourcing; and (2) minimizing the adverse effects of comments from unqualified individuals or internet trolls (i.e., individuals who appear when any online system gains notoriety and seek to disrupt the system put in place). Various systems are being developed to address these challenges, including efforts that integrate traditional with informal peer review in a collaborative process or enable crowdsourcing without the disruption from those with negative intent. For example, Alphabet Inc. has leveraged the technical capabilities of its subsidiary, Google, to create a searchable database of scholarly literature [54] that enables publication-sharing. Google Scholar maximizes publication access that leverages crowdsourcing through its publication ranking process. Google Scholar includes various criteria indicative of a document's value (e.g., citation rate, authorship, journal or publication service, and content) to rank documents [54]. Through this process, Google Scholar attempts to function symbiotically with scientific journals and established peer review practices.

Other attempts at crowdsourcing have taken a more granular approach. For example, researchers sought to evaluate whether crowdsourcing could be used to evaluate scientific literature. In the study, the researchers examined the distributed the literature evaluation work via Amazon's Mechanical Turk, which is a marketplace for the hiring of "microworkers," individuals that take up very specific, short term tasks that require human intelligence [2]. The researchers concluded that "with good reliability and low cost, crowdsourcing has potential to evaluate published literature in a cost-effective, quick, and reliable manner using existing, easily accessible resources" [15].

In another example, the chemistry journal, SYNLETT, organized over 100 qualified reviewers into a crowdsourcing review team to evaluate scientific manuscript submissions. They gave one review team 72 h to evaluate a submission and recorded the team's comments. Their results demonstrated that the crowdsourced review team "showed at least as much attention to fine details, including supportive information outside the main article, as did those from conventional reviewers" [81]. These examples reveal a potential for a system wherein large numbers of qualified individuals, who currently have not been involved in the scientific review process, are motivated and encouraged to participate.

2.4   Mass Adoption of Open Access Publication

Ultimately, developments in open access publication will be ineffectual if widespread adoption of open access policies and technologies are not widely accepted. The point at which mass adoption of open access platforms occurs depends on various factors, including as a condition of funding and through the advocacy of thought leaders. For example, in 2015, the Bill and Melinda Gates Foundation established an open access policy that applies to all publications describing research it funds. The policy states that all applicable publications must be discoverable and accessible online, published under a Creative Commons license, and accessible and open immediately. In addition, the policy requires all data underlying research results to be made accessible and available immediately after publication. The Gates Foundation attempts to tackle a major reservation by many opponents of the open access policy by including reasonable costs for publisher fees in awards [11]. The Foundation's support for open access may help catalyze other efforts relevant to this movement.

The ability to share scientific data quickly and easily may lead to numerous positive societal benefits but also could introduce new risks. Looking to the future, developing and implementing approaches for harnessing scientific advances while simultaneously minimizing risks will be necessary to ensure stakeholders reap the benefits of biotechnology without exploitation. This balance is the heart of the *dual-use* consideration relevant to the emergence of new technologies and biosecurity.

3   Cross-Disciplinary Science in Biological Research

Small and large changes in science and technology influence the advancement and creation of new tools, knowledge, and concepts in biology. Small changes generally occur as part of the normal cycle of fundamental and applied research wherein new discoveries lead to additional scientific questions or challenge established tenants. Larger changes tend to occur when new fields of science, disciplines, or conceptual frameworks converge. Notable examples include additive biomanufacturing, which applies the technologies of 3D printing with the development of "ink" made of living cells; synthetic biology systems (e.g., combining DNA-based "circuits" to create an organism of known function), which applies engineering and computer science problem-solving paradigm to the field of biology; and systems biology, which uses advanced data analytics and visualization to study connections and networks within and between molecules, pathways, cells, physiological systems, and life forms. In addition, transformative changes in society could result from trial-and-error during the discovery process or through careful observation of biological systems. Examples of these changes include genetic engineering, which was enabled by the discovery of proteins that cut DNA internally (called endonucleases), and CRISPR-based genome editing, which emerged from research on immune responses in bacteria. This section will explore these advances in biotechnology.

3.1   Artificial Intelligence in Biomedical Science

Recent breakthroughs in artificial intelligence (AI) and supportive analytical software is influencing biological research and development. Although AI has been in existence for more than fifty years, recent advances in machine learning, "big data", and the development of the computational infrastructure have resulted in the transformative application of AI to the life sciences [89]. A relatively well-known example of this application of AI to biological study is IBM's Watson. Watson is being used to analyze vast amounts of genomics research for drug discovery and patient care management [62]. Further, AI is being used to model and characterize biological mechanisms for cancer and other complex diseases. Instead of using animals, cells, or humans to study disease and medical interventions, AI and large-scale statistical analyses are increasingly being used to understand how the body changes and disease progresses in silico (i.e., computationally). These studies enable computational analysis of genomic data, changes in an organism caused by a stimulus or mutation, and anticipated effects of medical interventions (e.g., through the use of a computer-designed human body). These capabilities promote the "bottoms-up" design of cells or organisms that have specific functions without an existing genetic "blueprint" [104]. For example, AI-designed organisms could be developed to alter bioelectronic properties when exposed to a particular chemical, or to generate energy from an individual wearing the material while exercising.

This capability is much less "science fiction" and closer to "inevitable science future" than many would believe. Recognizing the continued blurring of lines between biology, engineering, and technology, the U.S. government established the Biological Technologies Office at the Defense Advanced Research Projects Agency (DARPA) in 2014 to "explore the increasingly dynamic intersection of biology and the physical sciences" [30]. Additionally, artificial neural networks (a machine learning technique inspired by biological neural networks), have demonstrated the ability to "learn" and self-improve through the computation of vast quantities of data [118]. These capabilities have led to breakthroughs in responsive prosthetics and robots that can sense environmental surroundings [34, 117]. These technological advances have the potential to address a variety of biological and broader national security risks through the development of biosensors, military medical applications, early biological, chemical, and radiological warning and detection.

3.2 Big Data Analytics and Biology

Computational and statistical analysis tools have been developed and applied to the study of biological systems to better understand natural processes, physiological responses to medicines, and other foreign particles (e.g., pollutants, chemicals, viruses, and bacteria), and determinants of disease. These analytic tools include machine learning; network, image, audio, and language processing; and other data analysis technologies to integrate diverse and numerous data and generate results. This field, commonly referred to as "data science," promises to improve studies on environmental ecosystems, animal and plant systems, infectious disease surveillance, and health systems [10].

Systems biology and-omics sciences were the first field in biology to harness the capabilities of the convergence of data science and life science. These fields involve the generation and analysis of vast amounts of data to study the roles, relationships, and functions of biological molecules within a system (e.g., cell, organism, or ecosystem). The-omics sciences include genomics, proteomics, metabolomics, transcriptomics, and many other terms that describe the generation of large amounts of data about specific types of molecules (e.g., DNA or proteins) or about all molecules in a system (e.g., physiological changes after a medication is administered) [61]. These fields are enabled by the development of technologies for detecting and capturing the molecules such as next-generation sequencers that can detect DNA,

RNA, and methyl groups on DNA (i.e., epigenetic patterns), and the development and sharing of computational algorithms for analyzing omics data. In 2013, the U.S. NIH launched its "Big Data to Knowledge" (BD2K) program to transform the way biology is studied by investing in the development of new technologies and code development for analyzing omics data [96]. Through this initiative, the NIH aims to "maximize and accelerate" the application of big data analytics to the life sciences to improve biomedical research. This program includes funding support for research, centers of excellence that enable collaboration [96], and the training of scientists and students [98]. In addition to the NIH, large companies are using big data and data science in their research and development activities. For example, IBM is partnering with academic institutions to support cancer research through its Watson program. Additionally, Bristol-Myers Squibb (BMS), one of the world's largest pharmaceutical companies, uses Amazon's big data cloud systems to conduct clinical trial simulations. This allows BMS to conduct trial simulations 98% faster than other systems, leading to more cost-effective clinical trials [3].

3.3 Synthetic Biology

Another exciting development in biological research in the 21st century is design-based biology, specifically in the development of organisms and novel molecules that can be created to perform specific functions by design. This development led to the emergence of the field of synthetic biology, which is a multidisciplinary field that applies an engineering approach to biology where biological materials or systems are designed and synthetically developed for research and/or functional applications. The experimental approaches used in synthetic biology are similar to traditional genetic engineering. However, the primary difference between synthetic biology and genetic engineering is the heavy reliance on engineering design concepts in synthetic biology. The early practitioners of synthetic biology were engineers and computer scientists, who leveraged the decreasing costs and increasing capabilities of DNA synthesis. The first product produced using synthetic biology was synthetic artemisinin; artemisinin is a natural, plant-based molecule used to treat malaria infection [114]. These early efforts led to the creation of several bioengineering research and educational programs in universities and the involvement of citizen scientists in pursuit of various biological sciences activities. In addition, synthetic biology has enabled the development of new methods for DNA assembly, which helped to catalyze capabilities for chemically synthesizing large genetic sequences. A few teams of scientists have used synthetic biology to create viruses and bacteria from scratch, which has led to significant concern among national security experts about the possibility for synthesis of restricted pathogens [7, 95, 131]. However, these advances also have led to the development of an industry that designs individual organisms as "molecular factories."

Today's scientists have the ability to use synthetic biology to address a breadth of societal problems, including food, the environment, and human health. Large agricultural companies are using synthetic biology to improve crops, in part making them resistant to natural pests and disease. Additionally, individuals have used synthetic biology to create food products in the laboratory. For example, many commercial entities (e.g., Memphis meats, Beyond Burger, Impossible Foods) have synthetically developed meat in an attempt to reduce greenhouse gas emissions by decreasing reliance on meat production via livestock. As another example, synthetically engineered algae are being studied as potential, environmentally sustainable biofuels. Finally, certain synthetic biologists (e.g., The Open Insulin Project) seek to produce medications more efficiently and at a lower cost than traditional pharmaceutical production methodologies.

For synthetic biology to be developed and applied to these and other societal needs, standardized principles and language have been developed. Synthetic biologists and industry partners alike have initiated efforts towards the development and maintenance of standard biological parts (i.e., Synthetic Biology Open Language). One of the primary venues for synthetic biology, the International Genetically Engineered Machine Competition (iGEM) and associated BioBricks Foundation, also have created a legal framework to allow for the sharing of standardized, interchangeable genetic parts ("BioBricks"). These and other efforts seeking to standardize design principles and "biological parts" may improve synthetic biology's capabilities for addressing societal needs, especially if empirical scientific knowledge is generated at the same time.

3.4 DNA Storage

Within the past several years, academic scientists and large software companies have begun exploring DNA as a medium for information storage [13, 22, 53]. The rationale provided for these efforts was to develop a stable medium that has the capacity to store large amounts of information for a longer duration than traditional silicone. DNA is considered a highly stable molecule that has the ability to survive hundreds of years without demonstrable degradation. This level of stability, combined with decreasing costs associated with DNA synthesis, sequencing, and storage, have led scientists to examine DNA as an information storage medium. Although these efforts are in the early research phase, they are driving collaboration between computer scientists and synthetic biologists to make DNA a physical tool to store information. In 2017, two significant developments were described. The first was a demonstration that video footage could be encoded in DNA and subsequently retrieved and viewed [120]. The second was the development of a "biological teleporter" that could transmit digital code to pint viruses without human intervention [1].

4 DIY Research

The practitioners of biotechnology have diversified significantly in recent years. Interest in creating programmable and/or synthetic biological systems have driven individuals with various backgrounds (e.g., computer scientists and engineers) into biology. As a result, new bioengineering and synthetic biology approaches and techniques have been developed to create standardized approaches and biological parts for the development of novel biological systems, organisms, and molecules. Simultaneously, scientists in seemingly dichotomous academic disciplines have decreased disciplinary barriers to working with biology. Material scientists are discovering new ways of manipulating biological components, resulting in new approaches to biomedical design (e.g., 3D printing of tissues and organs). Neuroscientists and computer scientists have developed responsive robots and neuroprosthetics. Computational biologists have applied advanced data analytics to integrate, analyze, and visualize biological and non-biological data to understand the environment, healthcare, and human disease.

In the mid-2000s, individuals who had minimal or no scientific training (dubbed amateur or citizen scientists) began playing with biology in their kitchens, bedrooms, and garages [75]. Initially, these individuals were considered bio-hobbyists, who conducted simple experiments such as extracting DNA from strawberries or their mouths using household items. Others in this community were biological artists using biological materials as a medium with which to create art. As this movement grew, citizen biological scientists sought to conduct more sophisticated activities, such as viewing proteins on a gel or introducing DNA into cells [48]. However, many citizen scientists have difficulties in obtaining laboratory equipment and materials from biotechnology and chemical supply companies because of the cost and access barriers. These difficulties have sent some entrepreneurial citizen scientists to design and build laboratory equipment that is less expensive than commercial products and accessible (via open-source distribution) to the greater DIYBio community. These entrepreneurial efforts have resulted in the emergence of new companies and a new market, demonstrating to others in the community that more sophisticated biological techniques and devices could be designed in an affordable way to improve experimental procedures through the use of robotics and automation [119, 121, 129].

In addition to the creation of new businesses, the DIYBio movement has led to the establishment of community biology laboratories. Approximately 42 such laboratories exist in the U.S. and 62 laboratories exist throughout the rest of the world. These laboratories provide a place where citizen biologists can conduct scientific activities in a community environment. These laboratories offer open bench space to provide members centralized access to equipment that may have available outside the community facilities. These community biological laboratories often obtain discarded material, equipment and devices, including lab benches, centrifuges, PCR machines, and more. The equipment often is previously-owned and is purchased via online suppliers, such as Ebay and Craigslist, or supplied through another avenue, such as a moving university laboratory. These laboratories generally restrict the types of scientific studies that can be conducted to biosafety level (BSL) 1, to ensure that all community members are not exposed to any harmful microorganisms. Some community laboratories, such as GenSpace in Brooklyn, NY, have an advisory group that reviews proposed studies to identify potential biosafety risks and to ensure that the science can be conducted safely at the lowest biosafety levels. Furthermore, these community laboratories are required to comply with local and national laws.

The rise of community laboratories has transformed the DIYBio and professional research landscape significantly. These laboratories provide space for citizen scientists and have become locations where professional scientists choose to conduct research that is not supported by their institutions or by established funders. Increasingly, professional and citizen scientists are conducting more sophisticated activities in these community laboratories, including genetic engineering and genetic analyses of plants, microbes in urban environments, and food [72, 86]. Although many of these projects remain at the lowest biosafety level, some have raised significant concern because they introduce the ability to push both biosafety and bioethical boundaries. For example, professional scientists working in a San Francisco-based community laboratory initiated a project on glowing plants [52]. To raise funds, these scientists promised to distribute the genetically modified plant seeds to supporters on a crowdfunding platform [58]. This effort raised significant concern among environmental groups about the broader effects of introducing the modified plants into the environment. However, at the time (2013), the glowing plants were not covered under the existing regulatory scheme for biotechnology in the U.S. [52]. In 2017, the U.S. government updated its coordinated framework for regulating biotechnology products [38]. Additionally, the community laboratory where the initial activities were conducted began questioning whether the project violated its safety practices [52]. Other scientific activities conducted in community laboratories have included genetic manipulation of probiotics to sense melamine, examination of the genes associated with familial diseases, and development of simple, inexpensive kits for detecting malaria [69]. This increased accessibility to biotechnology is not without its risks. For example, one "biohacker" (Josiah Zayner) attempted to increase his muscle growth by injecting himself with a self-made genome editing (CRISRP-based tools) [16, 77]. Although his attempts have not produced the desired physiological effects, Zayner believed that his efforts demonstrated that CRISPR-based genome editing tools are accessible to and feasible for anyone to use.

Concerns about "biohackers" emerged in 2004, after Steven Kurtz, a university professor and artist who used bacteria as an art medium, was arrested for possessing the bacteria in his home [21], a concern that was based on fears of bioterrorism in the aftermath of the 2001 attacks in the United States [138]. Ultimately, the case was dismissed in 2008 [110]. This incident led many in the citizen science community to distrust law enforcement. However, active engagement by members of the scientific (both professional and citizen scientists), policy, and federal law enforcement communities have led to positive interactions between the DIYBio community and the Federal Bureau of Investigation (FBI) [79]. These efforts have resulted in new opportunities for these individuals to share their vision, purpose, and scientific interests with the security community and FBI, with the hope of reducing fears about illicit activities conducted by citizen biologists. Topics initially focused on raising awareness about the activities conducted by the DIYBio community, highlighting its role in informal science education (i.e., improving scientific communication to make it more accessible to the lay individual). A leader in this effort is the New York City-based laboratory, GenSpace, which provides open lab access, high school programs, and classes to interested members of its community. Their offerings include CRISPR-based genome editing and neuroscience [49]. These community-initiated engagement efforts have improved biosafety considerations and decreased overall security risks by promoting scientific inquiry and enhancing communication and transparency at DIY and community biological laboratories.

5 Challenges

Biotechnology advances largely from the influx of new practitioners and funders both of whom enable multidisciplinary efforts, innovation, and breakthroughs in science. These changes may provide new opportunities for generating scientific information about the human body and response to its environment (including responses to infectious disease, hazardous chemicals, changing temperatures, diet, exercise, and many other things). As information accumulates, scientists and technologists are able to develop new tools for improving health, agriculture, the environment, energy, and other sectors. However, these changes also may exacerbate existing challenges, including reproducibility of research and high cost of research. In addition, the changes may result in new challenges, such as the increased blurring of lines between basic, applied, and commercial research, and national security risks involving the use of biotechnology to gain economic or competitive edge. This section explores some of these challenges, focusing on the current landscape of biotechnology and the life sciences, more generally.

5.1 Reproducibility in Science

Reproducibility is a core principle of the scientific method. It refers to the ability of experiments and analyses to be repeated by the same or different researchers and result in the same findings or conclusions as the original experiments. The inability to reproduce research signals problems in the original studies. In the biomedical field alone, approximately 50% of studies are not reproducible, resulting in an annual cost of $28B or more in the U.S. This high degree of irreproducibility can have significant consequences across the research and development process for new medicines and affect a diverse range of stakeholders [45]. For example, funders may question the responsible and effective use of their financial support for research and/or become discouraged about the lack of progress towards achieving the research goals [46].

Similarly, the public and policymakers may view irreproducibility problems as inefficient use of government funds. Researchers may have difficulty in obtaining financial
support for future projects or they may lose their existing funds. In addition, the repu-
tations of researchers, research institutions, and journals may be harmed, presenting
future challenges in acquiring talented researchers, funding, and/or publication in
high-impact journals.

Although concerns about irreproducibility of research have existed for many years,
in 2012, Amgen researchers declared that they could not reproduce results from 88%
of their "landmark cancer papers" [6]. This admission resulted in widespread concern
among the scientific community about the degree to which published research
can be repeated successfully. In response, the scientific community established
new programs and platforms to promote reproducibility. Many scientific journals,
including *Nature*, *Science*, PLOS, and the Elsevier journals [101, 109, 122, 139] now
require researchers to make their raw data available for anyone to reanalyze. In addition, some journals have established expert groups to review manuscripts containing
certain types of data. For example, the journal *Science* has expert statisticians with
whom they consult on any manuscript that includes statistical analysis. This group
enables more rigorous peer review of manuscripts describing research involving
statistical analysis. Beyond the traditional scientific journals, bioRχiv (the platform
for sharing unpublished preprint articles in the life sciences) provides researchers
with opportunities to debate the merits of published research [24]. For example,
in 2017, scientists published articles describing their criticisms of the publication
on precise editing of a harmful mutation in viable human embryos [36, 82, 125].
Other efforts include the establishment on online research communities for sharing
and developing experimental methodologies [25] and the maintenance of a database
for retracted articles [135]. Finally, in 2016, the National Institutes of Health (NIH)
issued its guidelines on rigor and transparency of research, which requires grant
applicants to evaluate the rigor of the scientific studies that are used to justify study
design and to include a plan for authenticating reagents and cell lines used in the
research, if funded [97]. Although these efforts play a clear role in promoting reproducibility of life science research, they also may promote reproducibility in related
scientific fields. For example, the NIH has a website on which source code used in
biomedical studies is posted [99].

Advancing biotechnology may exacerbate the challenges of reproducibility,
particularly as new reagents are developed and shared by non-life science practitioners, computational algorithms are relied on for data analysis, and new equipment
are adapted for using and/or producing biological materials. The iGEM competition attempted to develop a repository of standardized "biological parts" from
which teams could obtain plasmids with genes-of-interest for their projects [63].
However, the plasmids included in the repository were not necessarily validated
at the time of deposition and/or changed over time, both of which reduce reproducibility of the plasmids and require the need for troubleshooting of experiments.
Another example involves assessing reproducibility of analyses generated from data
analytics, visualization, and machine learning algorithms [51]. The simultaneous
development of new computational methods and code, and their application to the

study of biology, ecology, agricultural systems, and health systems may create a situation in which analytical results may significantly vary each time the analysis is conducted. As these computational methods become more complex, the potential is high for not reproducing results from big data analytics and not being able to identify the basis of irreproducibility. This convergence between computational and life sciences extends beyond data analytics to the computer-aided design and printing of biological materials. For nearly a decade, scientists actively have been studying and developing methods for creating animal tissues and organs from using three-dimensional printing [116]. In practice, three-dimensional printing of tissues and organs requires the addition of biological factors (in a laboratory setting) to produce a functional tissue and organ. This reliance on computer-aided design and printing *and* laboratory methods in tissue and organ printing suggests that few stakeholders have the skills to produce, let alone, reproduce the results. Furthermore, the methods primarily have been developed within the private sector, suggesting that some information that may be critical for assessing reproducibility may be protected by intellectual property and trade secrets. Similarly, tacit knowledge (i.e., scientific knowledge learned through training) contributes to challenges of reproducibility, which may limit scientific advancement and duplication by malicious actors (a double-edged situation). For example, individuals who have the skills to chemically synthesize Severe Acute Respiratory Syndrome (SARS) coronavirus work in or have gained their skills by working in or with a few laboratories [59, 130]. Similarly, replicating the chemical synthesis of poliovirus requires specialized skills that are taught through apprenticeship in a laboratory setting [134]. Together, these examples highlight how biotechnology may present new and different challenges for checking whether research results are accurate and reproducible, a situation that may present barriers to leveraging the benefits of science, but that also may limit malicious use of biotechnology.

5.2   Pace of Change

The rapid pace with which biotechnology is advancing and being applied to health, agriculture, environment, and other sectors challenges the established systems for innovation and regulation. These challenges are best exemplified by CRISPR-based genome editing tools.

5.2.1   CRISPR-Based Tools

Within the past seven years, CRISPR-based genome editing tools have transformed the fields of gene therapy and genetic engineering, enhancing capabilities for deleting, inserting, or replacing sequences at precise locations in a cell's genome. These tools involve a small RNA molecule that recognizes and binds specific sequences in the genome, and a protein (Cas9) that cuts DNA. Variations of CRISPR-based tools include (a) dysfunctional Cas9 components that cut one strand of DNA or block the DNA from being accessed by other proteins; (b) use of different DNA-cutting components each with its own cutting pattern and level of precision; (c) modified Cas9 components that function in response to stimulus, such as light or the presence of another protein; and (d) coupling of these tools with a protein that can repair DNA-breaks (specifically in bacterial cells). These and other variations of CRISPR-based tools have expanded the number of organisms that can be modified using this system, ranging from difficult-to-culture bacteria to whole organisms (including plants, animals, and humans). In addition, these tools have improved capabilities for functional genomics (i.e., the study of gene sequence and function), correction of disease-causing mutations in humans and animals, physical enhancement of traits (e.g., enhanced muscle mass and petite pigs and monkeys), and pest resistance in plants. To enable these capabilities, new gene delivery approaches have been (and are continuing to be) developed. Already, CRISPR-Cas9 coupled to viral gene therapy vectors are commercially available and being used to modify human cells in the laboratory. To date, CRISPR-based tools have not been introduced directly into people (the published examples involved older generations of genome editing tools [65, 103]). However, CRISPR-based tools have been used to modify viable human embryos, [12, 76] an application for which strong human health drivers and ethical concern exist.

CRISPR-based tools have enabled the development of a new technology, gene drives. Gene drives are designed to push specific traits through a population much faster than would occur naturally in organisms that have short generation times (i.e., insects and small rodents). The Bill and Melinda Gates Foundation is investing $35 million in research to use CRISPR-based gene drives in mosquitoes that transmit malaria to reduce the size of the mosquito population in affected areas, which the Foundation hopes will reduce the number of new malaria infections [108] At the same time, researchers from New Zealand, Australia, and Texas have developed a gene drive in mice with the intent of reducing the size of the invasive mouse population in New Zealand [113]. Concern over the broader environmental effects of gene drives led DARPA to establish a program—Safe Genes—to understand gene drives better, develop approaches for their safe, responsible, and predictable use for beneficial applications, and address health and security risks of accidental or deliberately harmful use [31].

5.2.2 Regulatory Landscape

Governments and scientists worldwide have struggled with the rapid pace with which genome editing is advancing and being applied. The more controversial applications primarily have involved modification of human embryos and gene drives. At one end of the spectrum are arguments for further development of these applications, citing benefits to human health, specifically the correction of disease-causing genetic sequences or reduction of infectious disease burden for certain pathogens [84]. At the opposite end of the spectrum are arguments against development of these applications, raising concerns about creating "designer babies" or adversely altering the environment [90]. Addressing these concerns and leveraging the benefits of these technological advances through national-level regulation is difficult, if not feasible, because countries throughout the world have robust biotechnology sectors, the practitioners are diverse including professional and citizen scientists whose activities may not be covered by the country's existing regulatory structure, and the sources of funding have diversified limiting the reach of regulation that is tied to government-funding. These, and likely other factors, dictate the relevance and reach of existing policies within and among nations to address any potential ethical, safety, security, and other concerns associated with emerging biotechnologies.

In 2015, the U.S. National Institutes of Health issued a statement informing researchers that no federal funds will be used for editing human embryos, referencing legislative prohibitions against the "creation of human embryos for research purposes or research in which human embryos are destroyed" [100]. However, genome editing of human embryos was approved in Sweden, China, and the United Kingdom, indicating continued advancement of the methodologies and tools used to modify human embryos irrespective of the U.S.' stance [20]. Furthermore, in 2017, a research group based in the U.S. published its work on editing of viable human embryos [76, 82, 124, 125]. This work was funded entirely by non-U.S. government funds, which meant that the NIH and legal prohibitions on using federal funds to support research with human embryos did not apply. This reliance on federal funding to promulgate regulations will continue to present challenges to the U.S. system for governing research involving emerging biotechnologies because more 50% of U.S. research activities are not funded by the U.S. government [88]. In addition, after publication, the authors used the recently released U.S. National Academies report on ethics of human genome editing as justification for the study [67].

This CRISPR-Cas9 example highlights several challenges associated with research, development, application, and regulation of emerging biotechnologies. The pace with which these technologies are changing and being applied along with the increased diversity of practitioners, stakeholders, and funders in the U.S. and internationally will present hurdles for regulation, identification and promulgation of norms, and scientific research and development.

5.3   The Basic, Applied, and Commercial Research and Development Process

The research and development process for biotechnology products follows a general process of basic research, application to different problems, advanced development and testing, and final approval for commercial use. The process for clinical products include additional steps, including pre-clinical animal testing and three phases of clinical trials in humans to assess safety and efficacy of the products. Advanced development of clinical products occurs in Phase III trials, during which researchers gather data from hundreds of thousands of volunteers about the product's effectiveness at treating or preventing disease. If sufficient numbers of volunteers are not available, product developers may test product efficacy in animal models [43].

This basic process for product development may be changing over the next several years because of new technology developments. For example, the three-dimensional printing of tissues and organs are being used to create laboratory models for testing new product candidates that affect specific functions of organ model systems [107]. The primary drivers for these efforts are: (1) decreasing or eliminating the need for testing products in animal models and human subjects; (2) reducing the costs for early and advanced product testing; and (3) identifying products that may fail in advanced development at a much earlier step of the product development process. Reduction in experimental costs and early elimination of ineffective products ultimately can lead to lower overall product development costs.

Another advance that may change the product development process is the creation of computational modeling software that allows researcher to design and iterate on products before experimental laboratory work begins. This capability has the potential for reducing initial research costs because scientists can develop products through rationale design. However, synthetic biology has demonstrated that significant technical challenges exist in translating computational design to tangible product without trial-and-error [94, 112]. These challenges may not apply to iterating on a well-characterized product, where the structure and function in different formulations and under various conditions are known. In addition, the U.S. government, computational biologists, and synthetic biologists are attempting to describe and predict function from genetic sequence. Complicating these efforts is new research on non-genetically encoded determinants of function and traits. The development of computational analysis software and reliable scientific information could lead to greater adoption of computational design in early development of biological products.

New bioprocessing technologies are beginning to change the late-stage production and manufacturing of biological products, such as vaccine, bio-based therapeutics (e.g., antibodies), and altered cells. For more than 50 years, pharmaceutical companies have used a stepwise process called batch processing to make large amounts of biological products. Although reliable, this process can be slow, present challenges in converting facilities to produce new products, and be vulnerable to contamination in the in-between steps. However, within the past five years, the biotechnology industry has begun exploring a new processing method for making large amounts of biological products. Unlike batch processing, this new method, which is called continuous processing, involves the continuous, uninterrupted flow of materials from one step to another. Continuous processing can make product faster, reduce introduction of contaminants mid-production, increase productivity, and reduce overall waste compared to batch processing [50, 83]. Although some large pharmaceutical companies have begun using continuous manufacturing, its widespread use will be determined by a variety factors, including the level of uncertainty of the regulatory process, feasibility of converting existing facilities, and the facility conversion and ongoing manufacturing costs [92].

Beyond these biotechnologies, involvement of citizen scientists and the DIYBio community in more sophisticated life sciences activities also may affect the product development process. The iGEM competition has provided a platform for generating new, commercially relevant information and/or products. In 2011, a team from the University of Washington developed a system for producing a protein that degrades gluten in food and was several thousand times more effective than a similar product in industrial development [40]. The team, which consisted of undergraduate students and graduate-student advisors, established a start-up company (PvP Biologics, Inc) focused on further development of the product they created, KumaMax. In early 2017, PvP Biologics and Takeda Pharmaceutical Company Limited entered into an agreement for advanced development of KumaMax [17]. The initial investment in the iGEM project likely was $25,000, the minimum amount of funds teams are expected raise to complete their projects and compete in the competition. From start to finish, iGEM projects generally are conducted in a few months. This financial and time investment for involvement in the iGEM competition is significantly less than those typically involved in basic research and development, providing an incentive to monitor team projects for commercially relevant products.

The estimated cost of developing biotechnology and pharmaceutical products is over one billion dollars and can take more than 10-years [80, 91, 126]. Any advances—technological or process improvement—may reduce existing challenges in discovering, developing, and commercializing new biotechnology and pharmaceutical products.

5.4 Cost of Research

The relationship between research costs and biotechnology is challenging to capture, in large part because some advances decrease costs while others increases costs. Examples of technologies that have led to decreased research costs include: (a) advances in DNA sequencing methods and technologies that has resulted in an exponential decrease in the cost of sequencing [132]; (b) the expansion of the kits standardizing common molecular biology experimental procedures; (c) the emergence of supply companies selling used or older generation laboratory equipment[1]; (d) emergence of commercial services that sell custom-made viruses, plasmids, and other products; and (e) the integration of computer science with life-science activities (e.g., laboratory automation and cloud-based data analytics platforms). These advances have the potential to decrease the costs of equipping laboratories, conducting molecular biology experiments, and purchasing (rather than making) research materials. Some of these advances have enabled scientific activities in community laboratories.

---

[1] ALT. American Laboratory Trading. In; BioSurplus. In; Ebay. Lab Equipment. In; EquipNet. Used Lab Equipment. In; LabX. Biotechnology Listings. In; Cambridge Scientific. In; Biotech Equipment Sales. In.

Other biotechnology advances have increased the cost of research. For example, systems biology, multi-omic studies, and big data analytics have significantly increased research costs. These fields require advanced capabilities for computing and data storage (either in-house or via the cloud) to support the generation, analysis, and storage of exabyte (1×10^{18}) amounts of data, [78] which requires significant computational and physical infrastructure costing scientific institutions millions of dollars annually. In 2016, the National Institutes of Health was spending more $110 million a year to support the infrastructure and maintenance of 50 of its largest databases and the Gene Ontology Consortium was spending $3.7 million a year [78]. Other biotechnologies, such as three-dimensional printing of tissues, organs, and devices (e.g., microfluidic devices) and genome editing of mammalian cells, require specialized equipment, which can be expensive.

At the institutional level, biological research requires significant cost to maintain laboratory facilities, biosafety and ethics review committees, environmental health and safety activities, export control compliance, and technology transfer functions. In the U.S., these research costs are supported by indirect costs of awards [66] Research that involves compliance with numerous or highly complex regulations, large numbers of support staff, significant laboratory facilities, and/or significant computational and data storage needs have higher overall research costs than research that does not involve these features. Indirect cost rates to support these functions varies between 36 and 78%, with an average of 54.5% according to a 2010 survey of 72 research institutions in the U.S. [64].

The relationship between biotechnology advances and research costs is complex as described above. Biotechnologies that require specialized infrastructure or are used in highly regulated fields (e.g., pharmaceutical products, cell-based therapies, and Biological Select Agents and Toxins) will have high total costs of research. Conversely, biotechnologies that reduce the laboratory time and costs may lower the total cost of research that is not highly regulated. But, if used in highly regulated research, these biotechnologies do not necessarily lower the total cost of research.

5.5 New Security Challenges

Emerging biotechnologies challenges current concepts of biosecurity. Historically, biosecurity concerns have focused on the theft, diversion, and deliberate use of pathogens and toxins by malicious individuals to cause harm. Some biotechnologies, such as synthetic genomics may affect the ability of malicious individuals to acquire harmful pathogens through chemical synthesis, commercial services, or laboratories engaged in synthetic biology. However, most of the biotechnologies described in this chapter challenge this conceptualization of biosecurity, in large part because they do not involve the direct acquisition, modification, or use of pathogens and toxins. Rather, emerging biotechnologies may be associated with other national security risks, including economic, commercial, and health security risks. The challenge is identifying which characteristics and applications of biotechnologies contributes to these risks.

One characteristic of the current biotechnology landscape is the increased accessibility of biological tools, laboratory facilities, and information to a diverse array of practitioners. This increased accessibility promotes informal education, increasing scientific familiarity and literacy in the broader population, and enables stakeholders to know about, if not reap, the benefits of research. At the same time, the increases in accessibility of biology has raised concern among security experts who view it is lowering the barriers of entry for conducting biological research. Their arguments are that increased accessibility of certain information and biotechnologies could be exploited by malicious actors to cause harm. Furthermore, they argue that technological advances that simplify experiments, automate procedures, and/or leap-frog steps also could lower the barriers for malicious actors to exploit biology. Several security experts equate the trends in accessibility of biotechnology to that of computer science [56]. However, the issues of tacit knowledge (i.e., learned skills), technical capabilities, and financial and infrastructure resources of malicious actors are important considerations in understanding whether and the extent to which biotechnologies actually may lower the technological barriers.

One emerging risk that repeatedly presents concern among the national security community is the exploitation of biological data, but is not necessarily included in current dual use policies. The dual use (or, multi-use) nature of biotechnology became a significant policy issue in the U.S. and internationally in the early 21st century, particularly following the publication of the chemical synthesis of poliovirus and IL-4 mousepox virus [28]. In response to these concerns, the U.S. established the National Science Advisory Board for Biosecurity (NSABB) under the auspices of the NIH. The NSABB provided the U.S. government recommendations on criteria, review, oversight, and communication of dual use life sciences research of concern. In addition, the NSABB provided recommendations to prevent potential biosecurity risks of synthesis of Biological Select Agents in Toxins. These recommendations led to the development of the U.S. Screening Framework Guidance for Providers of Synthetic Double-Stranded DNA, and the 2012 and 2014 U.S. policies on oversight of dual use life sciences research of concern. Consistent with these, the U.S. government released a new policy guidance for care and oversight of research involving pandemic potential pathogens. These policies focus on pathogens and specific experiments or traits that could present risks to public health and safety.

None of these policies address potential exploitation of biological data that is not associated with pathogens. Yet, concerns about exploitation of genomic data has become a major concern among security experts, including the FBI [10, 35, 71]. These concerns are exacerbated by questions about the accessibility of available information, data analysis technologies, and platforms for cloud storage and computing. As big data analysis services become more prevalent, questions arise about whether they can be used by individuals with little formal education and training in data science or biology. In addition, security experts have expressed concern about actors exploiting cloud-based machine learning tools to evaluate vast amounts of scientific literature to glean useful information. These activities could shift the risk of biotechnology to the design phase or to the bioeconomy. Technical barriers, including tacit knowledge and scientific capabilities of actors, determine whether computer-based designs can be translated into a tangible product. However, malicious actors that are well-resourced and have the ability to obtain and analyze data may have little barriers to exploiting biological data to gain commercial, economic, or other advantage.

In addition, the increasing number of automated laboratory facilities and contract research organizations may be vulnerable to cyberattacks. The frequency with which computer networks are attacked and compromised highlights the need to integrate cyber security into physical and personnel security practices at biological research and production entities. Efforts have begun to identify and counter potential vulnerabilities at the intersection between computer and biological sciences, which may include the development of tools and policies that uniquely address biosecurity risks resulting from these vulnerabilities. Some organizations such as the members of the International Gene Synthesis Consortium (IGSC) are already beginning to address these risks, but more can be done to prevent unauthorized access to and exploitation of biological information.

The risks described in this section are among the more prevalent issues raised in recent years. As new information, technology capabilities, and applications are described, new and different risks may be revealed. Assessing these risks in the absence of technical feasibility and potential benefit to global health security, health, and other societal needs could result in misplaced or miscalculated concern. Therefore, a comprehensive approach for identifying and analyzing the risks and benefits of emerging biotechnologies could simultaneously prevent technological surprise while leveraging new capabilities to address critical health security and national security needs.

6 Practical Contributions of Biological Research to Global Health Security and National Security

Several of the advances mentioned in the preceding sections have contributed to the detection and response of natural and man-made biological threats. For nearly two decades, governments and private companies have supported the development of data integration and application platforms to gain early warning about potential outbreaks that could result in international public health emergencies. The earliest platform, ProMED-mail, was created in 1994 to enable scientists and public health practitioners share information about observed or suspected infections through email. More recently, ProMED-mail has developed a web-based interface that allows members to share information about new infections and learn about conferences and courses [111]. A few years after ProMED-mail was launched, the Canadian government created the Global Public Health Intelligence Network (GPHIN) to monitor media reports and other unofficial, online sources for identifying, compiling, and sharing information about possible infections [47, 136]. GPHIN was among the first platforms to use data analytics, specifically text analysis, to identify possible outbreaks [93]. These platforms continue to assist governmental, non-governmental, and intergovernmental stakeholders in early detection of natural, accidental, or deliberate biological incidents, a concept now referred to as biosurveillance.

Following the terrorist events of 2001, the U.S. became especially interested in building the scientific and technological capabilities for biosurveillance. The U.S. Centers for Disease Control and Prevention (CDC), Department of Homeland Security, and the U.S. Department of Defense DoD supported the development of systems to generate, integrate, and analyze a variety of data that could signal a potential biological incident. For example, the Department of Defense supported the development of ESSENCE (Electronic Surveillance System for the Early Notification of Community-based Epidemics), which is a system that integrates health and health-related data, including documented cases of people with specific conditions suggestive of infectious disease, pharmaceutical sales, observations of symptoms associated with different pathogens, clinical results, and information about reportable diseases [44]. This system sought to detect abnormal trends in the data that may signal the emergence of a devastating natural or man-made outbreak. The CDC developed BioSense to detect and assess potential bioterrorism incidents using data from state and local public health departments [55] and the Early Warning Infectious Disease Surveillance system to facilitate detection of transboundary diseases in North America [19]. The Department of Homeland Security established the National Biosurveillance Integration Center that integrates and analyzes health data from animals, plants, humans, and the environment to facilitate early warning of potential biological incidents [32]. Most recently, the Department of Defense supported the development of the Biosurveillance Ecosystem, which is a cloud-based platform that incorporates data-specific applications enabling users to detect potential anomalies in human and animal infections and prevalence [106]. These systems continue to take advantage of advances in cloud-computing and technologies for data integration, analysis, and visualization to enhance U.S. and international capabilities for detecting pathogens of national or international concern.

In addition to biosurveillance capabilities, the U.S. government has invested in the development of medical countermeasures (MCM), specifically vaccines and medicines, against high-risk pathogens that present significant public health and safety risks to the U.S. population and military. The Departments of Defense and Health and Human Services (HHS) have established research programs to support the development, testing, and acquisition of needed MCM. For example, DoD supported the construction of advanced manufacturing facilities for developing and producing biological products, one of which focused on medical countermeasures. This DoD Medical Countermeasures Advanced Development and Manufacturing (ADM-MCM) facility has state-of-the-art equipment and production capabilities to promote the development of "safe, effective, and innovative MCM" against chemical, biological, radiological, and nuclear agents. DoD has expressed interest in establishing public-private partnerships to increase opportunities for leveraging new technologies [8]. With similar intent, HHS has established a technology watch process for learning about new technologies that may improve MCM products and/or the MCM manufacturing platforms. Through this process, scientists have opportunities to introduce HHS to new biotechnologies or scientific advances that could be leveraged for MCM development [123]. In addition, the Food and Drug Administration [41] and NIH have invested in research to improve regulatory science, which involves the development of new approaches, standards, and technologies to assess the safety and efficacy of MCM [42]. These programs promote fundamental and innovative scientific initiatives to generate the data and tools needed to assess the safety and efficacy of vaccines and medicines that cannot be tested in people [42]. In addition, regulatory science investments seek to identify new processes and technologies that improve production time and scalability of MCM.

Emerging biotechnologies also has been explored for development of biosensors of chemical and biological agents [127]. Biosensors involve the use of living organisms or biological materials to detect agents or their signatures. Microbes, plants or other organisms naturally have specialized capabilities or could be engineered to detect molecular, enzymatic, electrical, and other signals [85]. Advances in applying biology for signal detection has prompted the U.S. government to invest in research and development of biosensors against chemical and biological threat agents. For example, DoD has expressed interest in developing biosensors for chemical and biological agent detection that are small, quick, highly sensitive, and highly accurate [102]. More recently, DoD has begun evaluating biosensor capabilities to determine whether they can be used to detect radiological or nuclear agents [18].

Emerging biotechnologies have been developed and applied to specific national security objectives such as situational awareness, incident detection, or response. However, a broader diversity of life-science research (including molecular biology, genetics, microbiology, immunology, virology, biochemistry, evolutionary biology, computational biology, biophysics, and bioengineering) can contribute to many more objectives, including the development of innovative approaches for:

- microbial forensics
- biosafety (e.g., engineered organisms that cannot survive in the environment)
- laboratory biosecurity (e.g., biometrics for identify verification)
- personnel security (e.g., behavioral studies of insider threat)
- military health (e.g., genomic and personalized medicine)
- pathogen characterization (e.g., study of host-pathogen response).

Biotechnology and life-science research have made several contributions to preventing, preparing for, and responding to natural and man-made biological threats [9]. In addition, biology can provide new opportunities for countering chemical, radiological, and nuclear threats, making it a versatile capability for promoting the health, safety and security of all people.

7 Summary

In 2000, Nobel Laureate Matthew Meselson declared the 21st century as the age of biotechnology at the annual meeting of the National Academies of Sciences. This statement came shortly after when the U.S. government and a private company (under the leadership of J. Craig Venter) was sequencing the human genome for the first time [115], academic scientists began studying cells and organisms as systems [70], the judicial system started using DNA identification methodologies as part of criminal investigations [26], and Scottish scientists announced they had cloned the first mammal, Dolly the sheep. These advances were enabled by decades of iterative improvements of the early genetic engineering methodologies, the rich body of knowledge generated from biological research, and the simultaneous advancement of computational capabilities. Biotechnology of the 20th century laid the foundation for 21st century biology and biotechnology.

Since 2001, the landscape for biotechnology has changed dramatically. The diversity of individuals who conduct biological research and develop innovative biotechnologies has grown. Computer scientists, engineers, physicists, chemists, materials scientists, and mathematicians began creating organisms as vessels for producing synthetic medicines, industrial chemicals, and compounds used in cosmetics. Their successes were enabled by advances in molecular biology and the continuous decreasing cost in DNA synthesis. These non-life scientists and engineers helped to establish and grow a new field, synthetic biology, which sought to enable individuals to design and create organisms with defined functions with a high degree of predictability and fidelity. Although this is an aspirational goal, scientists and engineers engaged in synthetic biology actively are working towards realizing it. A direct consequence of this goal was the creation of the iGEM competition, which has provided undergraduate students opportunities to develop innovative businesses and products that rival products in commercial development. Only time will tell what the participants in this competition and their mentors will achieve.

As these changes were occurring, citizen scientists interested in playing with and learning about biology began conducting simple, low risk activities, such as extracting DNA from strawberries or their own cells. The initial influx of amateur biologists, referred to ask biohackers or garage biologists, led to the creation of a new community of Do-It-Yourself biologists who eventually established community biolabs throughout the United States and internationally. Today, the DIYBio community has enabled new entrepreneurial efforts and opportunities to teach the public about biology. At the same time, this community has witnessed its members demonstrating cavalier attitudes by experimenting on themselves and selling products that may disrupt local ecology. This community has expressed their interest in promoting scientific awareness and familiarity in the general public, but this objective could be adversely affected if their members are viewed as careless or dangerous.

In addition to the increased diversity of practitioners, the funding landscape for biotechnology has changed. During the past twenty years, new models for financing life-sciences and biotechnology research emerged. Silicon Valley venture capitalists and crowd sourcing platforms have provided support to professional and citizen scientists that governments and philanthropic organizations will not fund. Some of these investments have supported the DIYBio community, while others have funded risky and unrealistic activities. Without government funding, these efforts likely fall outside the governance of biology and biotechnology. In addition to these private funders, many nations have begun supporting research in their countries to help address the challenges they face in health, agriculture, and environment. At least one country has expressed their interest in being a leader in biotechnology within the next two decades. This country, and potentially others, are leveraging scientific knowledge and skills outside their borders by providing full or partial funding to scientists in countries with more advanced capabilities in specific fields. This changing funding landscape has significant consequences on the ability of governments to regulate the development and application of biotechnologies to minimize harm or to harness the benefits of biotechnologies to address national needs.

The expansion of practitioners and funders has contributed to the convergence of science and engineering, leading to large changes in biotechnology capabilities and products. As the trend towards greater involvement in biology and cross-disciplinary activities continues, small and large changes in biology and biotechnology will take place. The iterative changes will expand capabilities of existing tools, as seen with CRISPR-based genome editing tools. The larger changes will add new, previously unseen applications, such as development of DNA as a data storage medium. The broader utility of new technologies and applications ultimately will rely on the technology itself, the regulatory environment, ease of use, availability of materials, and user need and acceptance. Until then, the full potential of emerging biotechnologies may not be realized. Underpinning the discussion of the changing face of biological research and progress is the need for biosecurity practices and applications to remain flexible.

References

1. Alexander B (2017) Biological teleporter could seed life through galaxy. MIT Technol Rev. https://www.technologyreview.com/s/608388/biological-teleporter-could-seed-life-through-galaxy/
2. Amazon. (n.d.). Amazon mechanical turk. https://www.mturk.com/worker/help
3. Amazon Web Services (n.d.). Bristol-Myers Squibb on AWS. https://aws.amazon.com/solutions/case-studies/bristol-myers-squibb/
4. ArXiv. (n.d.). About Arxicv. https://arxiv.org/help/general
5. Atkinson D (1998) Scientific discourse in sociohistorical context: the philosophical transactions of the royal society of London. Routledge, pp. 1675–1975
6. Baker M (2016) Biotech giant publishes failures to confirm high-profile science. Nature 530(7589):141
7. Baric R (2007) Synthetic viral genomics: risks and benefits for science and society
8. Belski T (2017) U.S. DoD opens advanced biologics manufacturing facility for private, public use. Bioprocess online. https://www.bioprocessonline.com/doc/u-s-dod-opens-advanced-biologics-manufacturing-facility-for-private-public-use-0001
9. Berger KM (2008) The role of science in preparedness and response. U St Thomas LJ 6:622

10. Berger KM, Roderick J (2014) National and transnational security implications of in the life sciences. American Association for the Advancement of Science, Washington, DC
11. Bill and Melinda Gates Foundation. Open access policy. https://www.gatesfoundation.org/How-We-Work/General-Information/Open-Access-Policy
12. Boddy J (2016) Swedish scientist edits DNA of human embryo. Science
13. Bornholt J, Lopez R, Carmean DM, Ceze L, Seelig G, Strauss K (2016) A DNA-based archival storage system. ACM SIGOPS Oper Syst Rev 50(2):637–649
14. Briggs H (2017) Fake research comes under scrutiny. BBC news. http://www.bbc.com/news/science-environment-39357819
15. Brown AW, Allison DB (2014) Using crowdsourcing to evaluate published scientific literature: methods and example. PLoS ONE 9(7):e100647
16. Brown KV (2017) Genetically engineering yourself sounds like a horrible idea - but this guy is doing it anyway. https://gizmodo.com/genetically-engineering-yourself-sounds-like-a-horrible-1820189351
17. Business Wire (2017) Takeda and PvP biologics announce development agreement around novel therapeutic for celiac disease. https://www.businesswire.com/news/home/20170105005656/en/Takeda-PvP-Biologics-Announce-Development-Agreement-Therapeutic
18. CBRNE Central (2017) New biosensor could help search for nuclear activity. https://cbrnecentral.com/new-biosensor-could-help-search-for-nuclear-activity/10601/
19. CDC (2004) Guidance for supplementary activities in support of the early warning infectious disease surveillance system. https://www.cdc.gov/phpr/documents/coopagreement-archive/fy2004/ewids-attachi.pdf
20. Callaway E (2016) Gene-editing research in human embryos gains momentum. Nature 532(7599):289
21. Center for Cognitive Liberty and Ethics (2004) FBI abducts artist, seizes art. http://www.cognitiveliberty.org/news/cae_arrest1.html
22. Church GM, Gao Y, Kosuri S (2012) Next-generation digital information storage in DNA. Science, 1226355
23. CiteSeerX (n.d.) CiteSeerX. http://citeseerx.ist.psu.edu/index;jsessionid=AB17C6470D17A30405FFB014100EA6D8
24. Cold Spring Harbor Laboratory (n.d.) BioRxiv. https://www.biorxiv.org/about-biorxiv
25. Collaboration OS (2015) Estimating the reproducibility of psychological science. Science 349(6251):aac4716
26. Cormier K, Calandro L, Reeder D (2005) Evolution of DNA evidence for crime solving–a judicial and legislative history. Forens Mag 2(4):1–3
27. Council NR (1982) Scientific communication and national security. National Academies Press, Washington, DC
28. Council NR (2004) Biotechnology research in an age of terrorism. National Academies Press
29. Creative Commons (n.d.) https://creativecommons.org/about/
30. DARPA (2014) DARPA launches biological technologies office. https://www.darpa.mil/news-events/2014-04-01
31. DARPA (2017) Building the safe genes toolkit. https://www.darpa.mil/news-events/2017-07-19
32. DHS (2010) National biosurveillance integration center strategic plan. https://www.hsdl.org/?abstract&did=767935
33. Demand Progress (n.d.) Demand progress. https://demandprogress.org/about/
34. Deplazes A, Huppenbauer M (2009) Synthetic organisms and living machines. Syst Synth Biol 3(1–4):55
35. Dusheck J (2015) Stanford researchers identify potential security hole in genomic data-sharing network. https://med.stanford.edu/news/all-news/2015/10/stanford-researchers-identify-potential-security-hole-in-genomic.html
36. Egli D, Zuccaro M, Kosicki M, Church G, Bradley A, Jasin M (2017) Inter-homologue repair in fertilized human eggs? BioRxiv 181255
37. Elife (n.d.) https://elifesciences.org/

38. Environmental Protection Agency (2017) Update to the coordinated framework for the regulation of biotechnology. https://www.epa.gov/regulation-biotechnology-under-tsca-and-fifra/update-coordinated-framework-regulation-biotechnology

39. Fanelli D (2009) How many scientists fabricate and falsify research? A systematic review and meta-analysis of survey data. PLoS ONE 4(5):e5738

40. Fell A (2017) From a student competition to a potential treatment for celiac disease. http://blogs.ucdavis.edu/egghead/2017/09/05/student-competition-potential-treatment-celiac-disease/

41. Food and Drug Administration (2018) Facilitate development of medical countermeasures to protect against threats to U.S. and Global Health and Security. https://www.fda.gov/ScienceResearch/SpecialTopics/RegulatoryScience/ucm268118.htm

42. Food and Drug Administration (2019) Advancing regulatory science. https://www.fda.gov/ScienceResearch/SpecialTopics/RegulatoryScience/default.htm

43. Food and Drug Administration (2002) New drug and biological drug products; evidence needed to demonstrate effectiveness of new drugs when human efficacy studies are not ethical or feasible, 21 CFR 314 and 601 C.F.R

44. Foster V (2004) Essence - A DoD health indicatory surveillance system

45. Freedman LP, Cockburn IM, Simcoe TS (2015) The economics of reproducibility in preclinical research. PLoS Biol 13(6):e1002165

46. Freedman LP, Venugopalan G, Wisman R (2017) Reproducibility 2020: progress and priorities. F1000Research, 6

47. GPHIN (2017) Global public health intelligence network. https://gphin.canada.ca/cepr/aboutgphin-rmispenbref.jsp?language=en_CA

48. Garage Biology (2010) Nature 467, 634. https://doi.org/10.1038/467634a

49. GenSpace (n.d.). https://www.genspace.org/

50. General Kinematics (2017) Batch versus continuous pharmaceutical manufacturing. https://www.generalkinematics.com/blog/batch-vs-continuous-pharmaceutical-manufacturing/

51. Gesing S, Connor TR, Taylor I (2015) Genomics and biological: facing current and future challenges around data and software sharing and reproducibility. arxiv preprint arXiv:1511.02689

52. Ghorayshi A (2013) DIY SynBio Lab "BioCurious" questions glowing plant project propriety – east bay express article. SynBioWatch. http://www.synbiowatch.org/2013/08/diy-synbio-lab-biocurious-questions-glowing-plant-project-propriety-east-bay-express-article/

53. Goldman N, Bertone P, Chen S, Dessimoz C, LeProust EM, Sipos B, Birney E (2013) Towards practical, high-capacity, low-maintenance information storage in synthesized DNA. Nature 494(7435):77

54. Google (n.d.) Google scholar. https://scholar.google.com/intl/en/scholar/about.html

55. Gould DW, Walker D, Yoon PW (2017) The evolution of BioSense: lessons learned and future directions. Pub Health Rep 132(1_suppl):7S-11S

56. Graham J (2016) Meet Apple's youngest app developer. She's nine. USA today. https://www.usatoday.com/story/tech/2016/06/15/meet-apples-youngest-app-developer/85951908/

57. Group R-DW (2009) RDF site summary. http://web.resource.org/rss/1.0/spec

58. Grushkin D (2013) Glowing plants: crowdsourced genetic engineering project ignites controversy. Sci Am

59. Gryphon Scientific (2016) Risk and benefit analysis of gain of function research

60. Harmon JE, Gross AG (2007) The scientific literature: a guided tour. University of Chicago Press

61. Horgan RP, Kenny LC (2011) 'Omic'technologies: genomics, transcriptomics, proteomics and metabolomics. Obstetric Gynaecol 13(3):189–195

62. IBM (n.d.) IBM Watson health. https://www.ibm.com/watson/health/

63. IGEM (n.d.) Registry of standard biological parts. http://parts.igem.org/Main_Page

64. Johnston SC, Desmond-Hellmann S, Hauser S, Vermillion E, Mia N (2015) Predictors of negotiated NIH indirect rates at US institutions. PLoS ONE 10(3):e0121273

65. Kaiser J (2017) A human has been injected with gene-editing tools to cure his disabling disease. Here's what you need to know. Science

66. Kaiser J (2017) NIH overhead plan draws fire: American association for the advancement of science

67. Kaiser J (2017) US panel gives yellow light to human embryo editing. Science https://doi.org/10.1126/science.aal0750

68. Kassner M (2013) Aaron Swartz legacy lives on with New Yorker's Strongbox: how it works. TechRepublic. https://www.techrepublic.com/blog/it-security/aaron-swartz-legacy-lives-on-with-new-yorkers-strongbox-how-it-works/

69. Keulartz J, van den Belt H (2016) DIY-Bio–economic, epistemological and ethical implications and ambivalences. Life Sci, Soc Policy 12(1):7

70. Kitano H (2002) Systems biology: a brief overview. Science 295(5560):1662–1664

71. Kozminski KG (2015) Biosecurity in the age of: a conversation with the FBI. Mol Biol Cell 26(22):3894–3897

72. Landrain T, Meyer M, Perez AM, Sussan R (2013) Do-it-yourself biology: challenges and promises for an open science and technology movement. Syst Synth Biol 7(3):115–126

73. Larkin M (2013) How to use crowdfunding to support your research. Elsevier Connect. https://www.elsevier.com/connect/how-to-use-crowdfunding-to-support-your-research

74. Larsen PO, Von Ins M (2010) The rate of growth in scientific publication and the decline in coverage provided by Science Citation Index. Scientometrics 84(3):575–603

75. Ledford H (2010) Garage biotech: Life hackers. Nature News 467(7316):650–652

76. Ledford H (2017) fixes disease gene in viable human embryos. Nat News 548(7665):13

77. Lee SM (2017) This guy says he's the first. BuzzFeedNews. https://www.buzzfeed.com/stephaniemlee/this-biohacker-wants-to-edit-his-own-dna?utm_term=.vd43m7E16D#.ggmgNy9JnD

78. Leetaru K (2016) The high costs of hosting science's: the commercial cloud to the rescue? Forbes. https://www.forbes.com/sites/kalevleetaru/2016/01/03/the-high-costs-of-hosting-sciences-big-data-the-commercial-cloud-to-the-rescue/#2a9da3ad28c0

79. Lempinen EW (2011) FBI, AAAS collaborate on ambitious outreach to biotech researchers and DIY biologists. AAAS News 1:2011

80. Letter, W. D. Biotechnology Development Costs Top $1.2 Billion Per Product. *FDA News*. https://www.fdanews.com/articles/93795-biotechnology-development-costs-top-1-2-billion-per-product?v=preview

81. List B (2017) Crowd-based peer review can be good and fast. Nature News 546(7656):9

82. Ma H, Marti-Gutierrez N, Park S-W, Wu J, Lee Y, Suzuki K, Ahmed R (2017) Correction of a pathogenic gene mutation in human embryos. Nature 548(7668):413–419

83. Massey S (2016) Making the switch: continuous manufacturing versus batch processing of pharmaceuticals. Xtalks. https://xtalks.com/continuous-and-batch-manufacturing-pharmaceuticals/

84. McFadden J (2016) Genetic editing is like playing God - and what's wrong with that? Guardian. https://www.theguardian.com/commentisfree/2016/feb/02/genetic-editing-playing-god-children-british-scientists-embryos-dna-diseases

85. Mehrotra P (2016) Biosensors and their applications–a review. J Oral Biol Craniof Res 6(2):153–159

86. Mello FN, Molino JVD, Alves TL, Ferreira-Camargo LS, Croce MA, Tanaka A, Carvalho JCM (2017) Overcoming adversities using DIYBIO and open science: open lab equipment and expression vector system validation in chlamydomonas reinhardtii. http://blogs.plos.org/collections/files/2017/04/USP-UNESP-UNIFESP-Original-Submission.pdf

87. Mervis J (2017). Data check: US government share of basic research funding falls below 50%. Science Magazine.

88. Mervis J (2017) Data check: US government share of basic research funding falls below 50%. Sci Mag

89. Morris KC, Schlenoff C, Srinivasan V (2017) Guest editorial a remarkable resurgence of and its impact on automation and autonomy. IEEE Trans Autom Sci Eng 14(2):407–409

90. Mukherjee S (2017) American scientists successfully edited faulty genes in human embryos. Is that ethical? Fortune. http://fortune.com/2017/08/03/gene-editing-human-embryos-crispr/
91. Mullin R (2014) Cost to develop new pharmaceutical drug now exceeds $2.5B. Sci Am https://www.scientificamerican.com/article/cost-to-develop-new-pharmaceutical-drug-now-exceeds-2-5b/
92. Munk M (2017) The industry's hesitation to adopt continuous bioprocessing: recommendations for deciding what, where, and when to implement. BioProcess Int. http://www.bioprocessintl.com/manufacturing/continuous-bioprocessing/industrys-hesitation-adopt-continuous-bioprocessing-recommendations-deciding-implement/
93. Mykhalovskiy E, Weir L (2006) The global public health intelligence network and early warning outbreak detection: a Canadian contribution to global public health. Can J Public Health/Revue Canadienne de Sante'e Publique, 42–44
94. National Institute of Standards and Technology (2016) Predictive engineering biology. https://www.nist.gov/programs-projects/predictive-engineering-biology
95. National Science Advisory Board for Biosecurity (2006) Addressing biosecurity concerns related to the synthesis of select agents. https://osp.od.nih.gov/wp-content/uploads/2013/12/Addressing%20Biosecurity%20Concerns%20Related%20to%20the%20Synthesis%20of%20Select%20Agents.pdf
96. National Institutes of Health (n.d.a) to knowledge. https://commonfund.nih.gov/bd2k
97. National Institutes of Health (2018) Rigor and reproducibility. https://grants.nih.gov/reproducibility/index.htm
98. National Institutes of Health (n.d) Training, education, and workforce development. https://datascience.nih.gov/bd2k/funded-programs/enhancing-training
99. National Institutes of Health (2006–2017) Source code for biology and medicine. https://www.ncbi.nlm.nih.gov/pmc/journals/449/
100. National Institutes of Health (2015) Statement on NIH funding of research using gene-editing technologies in human embryos. https://www.nih.gov/about-nih/who-we-are/nih-director/statements/statement-nih-funding-research-using-gene-editing-technologies-human-embryos
101. Nature (n.d.) Data availability statements - guidance for authors and editors. http://www.springer.com/gp/springernature/group/data-policy/data-availability-statements
102. Office of Technology Assessment (1984) Commercial biotechnology: an international analysis
103. Olena A (2017) First in vivo human to be tested in new clinical trial. Scientist. https://www.the-scientist.com/?articles.view/articleNo/49456/title/First-In-Vivo-Human-Genome-Editing-to-Be-Tested-in-New-Clinical-Trial/
104. Pasotti L, Zucca S (2014) Advances and computational tools towards predictable design in biological engineering. Computational and mathematical methods in medicine
105. Paulson K (2013) Strongbox and Aaron Swartz. The New Yorker. https://www.newyorker.com/news/news-desk/strongbox-and-aaron-swartz
106. Pellerin C (2016) DTRA scientists develop cloud-based biosurveillance ecosystem. DoD News. https://www.defense.gov/News/Article/Article/681832/dtra-scientists-develop-cloud-based-biosurveillance-ecosystem/
107. Peng W, Datta P, Ayan B, Ozbolat V, Sosnoski D, Ozbolat IT (2017) 3D bioprinting for drug discovery and development in pharmaceutics. Acta Biomater 57:26–46
108. Philanthrophy News Digest (2016) Gates foundation awards $35 million for mosquito research. https://philanthropynewsdigest.org/news/gates-foundation-awards-35-million-for-mosquito-research
109. PloS (n.d) Data availability. http://journals.plos.org/plosmedicine/s/data-availability
110. Press TA (2008) Charge dropped against artist in terror case. The New York Times. http://www.nytimes.com/2008/04/22/nyregion/22bioart.html
111. ProMED. (Producer). (February 12, 2018). ProMED Mail. https://www.promedmail.org/supporters/
112. Purcell O, Jain B, Karr JR, Covert MW, Lu TK (2013) Towards a whole-cell modeling approach for synthetic biology. Chaos: Interdisci J Nonlinear Sci 23(2):025112

113. Regalado A (2017) First gene drive in mammals could aid vast New Zealand eradication plan. MIT Technol Rev. https://www.technologyreview.com/s/603533/first-gene-drive-in-mammals-could-aid-vast-new-zealand-eradication-plan/?mc_cid=d82dc3c1d1&mc_eid=1294cc 0457

114. Ro D-K, Paradise EM, Ouellet M, Fisher KJ, Newman KL, Ndungu JM, Kirby J (2006) Production of the antimalarial drug precursor artemisinic acid in engineered yeast. Nature 440(7086):940

115. Roberts L, Davenport RJ, Pennisi E, Marshall E (2001) A history of the Human Genome Project. Science 291(5507):1195

116. Roskos K, Stuiver I, Pentoney S, Presnell S (2015) Bioprinting: an industrial perspective.In: Essentials of 3D biofabrication and translation. Elsevier, pp. 395–411

117. Rossiter J (n.d) Robotics, smart materials, and their future impact for humans. Open-Mind. https://www.bbvaopenmind.com/en/article/robotics-smart-materials-and-their-future-impact-for-humans/?fullscreen=true

118. Ruano MG, Ruano AE (2013) On the use of artificial neural networks for biomedical applications. Soft computing applications. Springer, pp. 433–451

119. Saenz A (2010) Making the modern do-it-yourself biology laboratory (video). SingularityHub. https://singularityhub.com/2010/08/03/making-the-modern-do-it-yourself-biology-laboratory-video/#sm.00016x9sf4152jdvypy0i9ir2i8ru

120. Sample I (2017) Harvard scientists pioneer storage of video inside DNA. Guardian. https://www.theguardian.com/science/2017/jul/12/scientists-pioneer-a-new-revolution-in-biology-by-embeding-film-on-dna

121. Sarah Z (2016) Inside the robut-run genetics lab of tomorrow (Just watch your step. Wired. https://www.wired.com/2016/08/inside-robot-run-genetics-lab-tomorrow-just-watch-step/

122. Science. Sci J: Edit Policies. http://www.sciencemag.org/authors/science-journals-editorial-policies

123. Seals J (2014) Engaging with BARDA: TechWatch program

124. Servick K (2017). First U.S. team to gene-edit human embryos revealed. Science

125. Servick K (2017) Skepticism surfaces over CRISPR Human embryo editing claims. Science

126. Stewart JJ, Allison PN, Johnson RS (2001) Putting a price on biotechnology. Nat Biotechnol 19(9):813

127. Stratis-Cullum D, Sumner J (2010) Chapter 6: Biosensors and bioelectronics. In Armstrong RE, Drapeau MD (eds) Bio-Inspired innovation and national security. NDU Press

128. Swartz A (2016) Guerilla open acess manifesto. https://archive.org/details/GuerillaOpenAccessManifesto

129. Synthase (n.d) https://synthace.com/2014/11/14/synthace-announces-the-release-of-antha-the-open-source-language-for-biology/

130. Thao TTN, Labroussaa F, Ebert N, V'kovski P, Stalder H, Portmann J, . . . Kratzel A (2020) Rapid reconstruction of SARS-CoV-2 using a synthetic genomics platform. bioRxiv

131. Tucker JB (2011) Could terrorists exploit synthetic biology? The New Atlantis, 69–81

132. U.S. National Human Genome Research Institute (2016) The cost of sequencing a human genome. https://www.genome.gov/27565109/the-cost-of-sequencing-a-human-genome/

133. Van Noorden R (2014) Global scientific output doubles every nine years. Nature News Blog

134. Vogel KM (2012) Phantom menace or looming danger?: a new framework for assessing threats. JHU Press

135. Watch R (Producer). (n.d) http://retractionwatch.com/

136. World Health Organization (n.d) GPHIN (intelligence tool) and GOARN (operational arm) of IHR (2005)

137. Yang Y, Zhang Y (2017) China launches crackdown on academic fraud. Finan Times. https://www.ft.com/content/680ea354-5251-11e7-bfb8-997009366969

138. Younge G (2004) Art becomes the next suspect in America's 9/11 paranoia. Guardian. https://www.theguardian.com/world/2004/jun/11/arts.usa

139. Zudilova-Seinstra E, van Hensbergen K (2016) How new article types help make science more reproducible

Emerging Biosecurity Considerations at the Intersection of Biotechnology and Technology

Stephen M. Lewis

Abstract Humanity is at a unique point in history where what was once believed to be science fiction is now shifting to become the real, emerging technology of today. Tantamount to this development of new technologies is the convergence of previously dichotomous disciplines including computer science, cellular biology, molecular biology, mechanical engineering, and technical disciplines. At the intersection of biotechnology and technology are emerging biosecurity considerations worth exploring in the context of innovation, biological design, manufacturing, automation, and artificial intelligence. There are also biosecurity considerations related to individuals and new skill sets developing as a result of technological and biotechnological progress. This chapter explores the various intersections of the world's most advanced technologies and outlines a model for the "full stack biotechnologist," a multiskilled expert with myriad capabilities and new considerations for risk and threat potentials. As technology and innovation progress to new heights and capabilities, perhaps the magnum opus of silicone advancement dictated by Moore's Law will be its impact on giving rise to its potential replacement: biotechnology.

1 Design Principles and Rapid Prototyping Applied to Biotechnology

1.1 Biological Design

New technology paradigms follow a familiar pattern: an advanced technology with a world of applications is created by academic-, corporate-, or government-funded research and development (R&D); then, slowly, this advanced technology evolves over time to become cheaper and more accessible to a broader and more diverse group of interested end users. When technology is first developed, function and application

S. M. Lewis (✉)
Thermo Fisher Scientific, 387 N. Corona St. #575, Denver, CO 80218, US
e-mail: steve@lewis.media

© Springer Nature Switzerland AG 2021
R. N. Burnette (ed.), *Applied Biosecurity: Global Health, Biodefense, and Developing Technologies*, Advanced Sciences and Technologies for Security Applications,
https://doi.org/10.1007/978-3-030-69464-7_7

take precedence over form. However, as more individuals gain the requisite skill set to work with or develop their own use for the technology, new applications and design embodiments emerge. It is this cycle that allows for true innovation to occur through the process of prototyping. When design principles combine with the ability to explore through rapid iteration, paradigm shifting advancements turn into true revolutions. This is the state of biotechnology today.

1.2 Increasing Access, Decreasing Costs

Requisite to nearly every emerging paradigm shift is increased access to a new technology and decreased cost for experimentation. Decreasing costs allow traditionally expensive technologies to reach previously excluded innovators, thus increasing access. As this convergence has continued to occur over the 2000s and 2010s, advanced technologies have been built by innovators who otherwise would have been excluded. In recent years, this democratization is perhaps easiest to appreciate and illustrate at the intersection of electronics and the do-it-yourself (DIY) movement. In less than two decades, designing and experimentation with electronics went from:

1. Only large corporations, to
2. Hobbyists with a microcontroller, a breadboard, a deep level of technical expertise, and a copy of *Physical Computing* [24], to
3. The emergence of electronics companies such as *littleBits* and *Arduino*, making electronics and programming easier and more accessible to students in K-12 programs.

Biotechnology is following this pattern at a much faster rate. Globalization and the continued widespread adoption of the Internet have led to increased access to a nearly all biological technologies that were once cost prohibitive to experiment with as an individual, group, or small company.

1.3 Rapid Prototyping

As a collateral impact of increased access and decreased costs, open access to hardware and electronics has allowed citizen scientists to begin building their own laboratory equipment (e.g. PCR, centrifuges, micropipettes) to outfit their home and community laboratories. Combined with free and open-source software (FOSS), these amateur scientists are able to rapidly prototype laboratory equipment with little more than a 3D printer and a microcontroller, such as *Arduino* [19]. A particularly interesting aspect of this increased access is that users are able to alter and rapidly prototype their designs, often times resulting in equipment that is not only cheaper, but also higher quality and specialized to their exact needs. All of this is being done for a fraction of what it would cost to purchase from traditional equipment vendors and manufacturers, who have a significant amount of overhead associated with their business models.

1.4 Biosecurity Considerations

Tantamount to the convergence of previously dichotomous disciplines, increased access, improved processes, and decreased cost to biotechnology is an emergence of new threats, risks, and vulnerabilities from a biosecurity perspective. Increased access and the ability to rapidly prototype with biology is an emerging market sector that has its own unique biosafety and biosecurity considerations. Biosecurity practitioners are in the process of exploring the potential impacts associated with these new participants and approaches to experimentation in biotechnology as a result of this convergence.

2 The "Full Stack Biotechnologist"

Before a biosecurity practitioner can explore new threat potentials, it is perhaps best to explore the concept that a new breed of biotechnologist is emerging. This proposed model is someone with a diverse set of skills ranging from computer programming, to molecular biology, to design, to manufacturing. The proposed model is known as the "Full Stack Biotechnologist."

The concept of "full stack" is derived from software development, where a "full stack developer" is someone who can do it all: design, code, test, and build in a range of languages required for both front end (user-facing) and back end (database) software development. Front end development in today requires knowledge of a variety of languages (e.g., HTML, CSS, and Javascript) to make dynamic applications that users see and interact with. Back end development requires knowledge of databasing technologies (e.g., SQL, NoSQL). These full stack developers also have an ability to work with and design the application protocol (HTTP), allowing for interaction between front end and back end components. True "full stack" developers also have a holistic knowledge and ability to troubleshoot the overall architecture of a software product [11]. Perhaps the most obvious benefit of a full stack developer comes from the fact that they have the knowledge base of what were traditionally three or four discrete developer types.

Applied to biotechnology, the concept of "full stack" is interesting to consider as increased access and decreased cost allow individuals to learn an entire pipeline of biotechnology from genome engineering through bioprocessing. Never before in history has a single individual had the opportunity to learn such a diverse range of skills in the biological sciences. The model proposed in Table 1 draws a parallel between "Front end" as "downstream" and "Back end" as "upstream" skill sets.

Table 1 Skills of a full stack biotechnologist

Upstream ("Back end") skills	Downstream ("Front end") skills
Experience with recombinant DNA technologies (e.g., CRISPR, RNAi)	Purification
Cell culture	Filtration
Scale-up and bioprocessing	Marketing
Bioprocessing	Design
Knowledge of cell and molecular biology	Rapid prototyping
Programming	
Machine learning	

Of course, no model is entirely clean and perfect. The overall model above deserves further thought and adaptation for a specialized biosecurity assessment framework; however, just as the idea of a "full stack biotechnologist" is emerging, so too are the foundational ways in which to categorize their myriad skill sets. From a biosecurity perspective today, it is simply worth considering that never before in history have so many fields converged toward such a broad potential for innovation from an individual. From the biosecurity perspective, this represents a shift in the application of biosecurity from an enterprise-level function toward impetus on individual-level accountabilities.

3   New Approaches to Biological Design and Development

A recurring theme in the context of biosecurity is the concern that many new tools and technologies are available to individuals, as explained above in the full stack biotechnologist. The idea that biology gone "unchecked" is a potential threat to the well-being of human, animal, and environmental health, or through the lens of biodefense, is one that challenges academics and policy makers alike. The other side of this, of course, is the intellectual capital that results from the dispersal of these technologies to a wider array of innovators. Examples of these include the design and development stages in the realm of biology.

3.1   Design of Novel and Existing Organisms, Biomolecules, and Biotherapeutics

A critical topic of exploration for emerging biosecurity considerations at the intersection of biotechnology and technology is the development of novel and existing organisms, molecules, and therapeutics through non-traditional mechanisms.

In 2015, Anthony Di Franco started a project at the community bio lab, Counter Culture Labs, in California to make insulin for type 1 diabetics for a significantly decreased cost, compared to traditional pharmaceutical prices [9]. Their project has been met with skepticism, fear-mongering, and concerns over patent infringement. While these concerns may be warranted under certain lenses, the motivation for the project is primarily altruistic, rooted in the concept that access to insulin is a fundamental human right. Whether this concept is right or wrong, this project has elicited significant conversation surrounding what it means to develop biologics through individual or group efforts, despite hesitation and lack of regulatory support. It is worth noting that this particular community bio lab champions biosafety and biosecurity; however, copycat projects based on this Open Insulin Project may not share their sense of altruism or security awareness.

Another example includes the emergence and success of the synthetic biology company, Ginkgo Bioworks [12]. Their success is built around their modular, flexible, and scalable platform for the design and scale-up of organisms for the production of biomolecules. The platform they have built is organism and end product agnostic, which allows for the company to weave a culture of rapid prototyping and experimentation into their corporate fabric. Ginkgo Bioworks employs a diverse range of skillsets and contains perhaps the largest group of "full stack biotechnologists" in the emerging synthetic biology market. As a result, they have the ability to design, build, and test novel organisms and metabolic pathways throughout their product development pipeline.

Another example, and potentially an initial catalyst for federal interest in novel organism development from a biosecurity perspective is the fact that poliovirus was synthesized in 2002 by a team that assembled a complementary DNA (cDNA) sequences of the poliovirus genome sourced from open access gene repositories. The cDNA was used to produce viral RNA, and subsequently used to produce infectious poliovirus virions [4]. This team was primarily driven by scientific motivations; however, it served as a proof of concept for a potential bioterrorism pipeline. After this journal article was published, the U.S. government became further interested in this topic area and has developed additional programs of research that are discussed further in this chapter.

3.2 Development: Wetware, Software, Hardware, and Manufacturing

With design and rapid prototyping cycles shifting the approach to experimentation with biotechnology, so too are new forms of wetware, software, hardware, and manufacturing tools and techniques emerging to support this experimentation. For example, in recent years, there has been a significant increase in experimentation with microfluidic devices. Although there has been a rekindled interest in microfluidics, they were first rapidly prototyped in 1998, when polydimethylsiloxane microfluidic devices were first designed to separate out amino acids and DNA fragments in aqueous solutions, with resolution comparable to fused silica capillaries [8]. As a true testament to the recently reignited popularity of microfluidic devices, this paramount journal article was released nearly 15 years before it gained significant notoriety.

Additionally, although microscopy has been a relatively static field for some time, this changed in 2015 when a group of scientists from MIT invented the field of expansion microscopy. These researchers synthesized a swellable polymer network within a specimen, to physically expand it. This resulted in physical magnification and the ability to perform scalable super-resolution microscopy with diffraction-limited microscopes [5]. The most impressive aspect of what this group accomplished is the accessibility and cost at which they did it. These researchers have devised a mechanism by which expansion microscopy can be applied to the preparation of human clinical specimens using readily available reagents for a few dollars per specimen. Tools like expansion microscopy will certainly enable further discovery and development of more advanced technologies, at an affordable cost for others to utilize and build upon.

From a hardware perspective, there are a considerable number of new projects and innovations developed recently. One such innovation is a science, technology, engineering, art, and math (STEAM) kit, developed by Amino Labs, designed to make bioengineering more accessible to students in K-12 programs. Another recent invention includes the OpenTrons OT-2, which was originally funded as a Kickstarter campaign. The OT-2 is a liquid handling robot and protocol and scientific workflow standardization platform designed to make science less labor-intensive, more efficient, and more reproducible for all. Hardware technologies are increasingly emerging as combination software platforms to further serve the advancement of science and to lower barriers for entry. This has significant benefits for increased access and participation, and it is important to consider the intentions behind the use of these technologies as dual-use, however minor.

Additionally, advances in technologies related to industrial-scale manufacturing are occurring at an impressive pace around the world. Once thought impossible, 3D metal printing is more and more emerging as a practical means to supplement traditional metal casting and CNC milling [15]. Additive manufacturing is the addition and joining of materials to create something new. The most popular form of additive manufacturing to date is 3D printing [16]. Biotechnologists are applying this concept to biological molecules and are also having success [13]. It is reasonable to imagine as advanced manufacturing techniques are further developed, practitioners of biotechnology will be inspired by best practices for the development of biomolecules and scale up of organisms.

Distributed manufacturing is another emerging approach whereby components of a larger design are developed on-demand in distributed locations for the purposes of assembly [14]. One can imagine with the near-simultaneous emergence of cloud laboratories (e.g., Transcriptic, Emerald Cloud Lab) that distributed manufacturing is an approach that could be applied to the development of synthetic molecules, organisms, or cell-free systems at scale. Advanced manufacturing applied to biotechnology can present new biosecurity considerations, as distributed manufacturing is essentially what occurred with the recombinant poliovirus example described above.

Related to the concept of manufacturing is the improvement of bioprocessing and scale-up technologies. Advancements in bioprocessing has given rise to innovative new companies developing novel approaches for scaling up the production of biological materials for the development of consumer products. Recent success stories include cell culture-based meat companies (e.g., Impossible Foods, Memphis Meats), yeast fermented recombinant spider silk designer clothing companies (Bolt Threads), and mycelium-based materials design (Ecovative) companies [1]. The greatest challenge of biotechnology to date has been the ability to manufacture a desired product or organism at scale, ensuring quality assurance and quality control along the way. This is the reason why pharmaceutical companies are recently producing more monoclonal antibody biotherapeutics (e.g., adalimumab) than ever before in history.

4 The Intersection of Artificial Intelligence and Biotechnology

In recent years there have been impressive advances in genomics, signal transduction pathway analysis, and protein folding using distributed computer processing techniques. Similarly, diagnostics, drug discovery, and biomarker identification are all being improved upon by the effective reemergence and advancement of machine and deep learning techniques and applications.

Uniquely applied for biosecurity purposes, these advances in bioinformatics are starting to emerge in the form of national security programs targeted at the identification of novel pathogens and potentially pathogenic nucleotide sequences. For example, The Defense Advanced Research Project Agency (DARPA) developed the Safe Genes program in 2017 to develop technologies to protect from the accidental or intentional misuse of genome editing technologies. The goal was to develop tools that can control, counter, and even reverse the effects of genome editing [28].

Another example in the scope of biosecurity includes the Functional Genomic and Computational Assessment of Threats (Fun GCAT) program of the Intelligence Advanced Research Project Agency (IARPA). The purpose of this program is to build tools the enhance the ability to functionally and computationally analyze nucleic acid sequences in order to ascribe threat potential to known and unknown genes. This would occur through computationally-derived comparisons to the functions of known threats and the ability to screen and identify sequences of concern. This effort further includes the identification of genes responsible for the pathogenesis and virulence of viral threats, bacterial threats, and toxins [10].

There are other programs with similar biosecurity and national security focuses and it is important to note that the only way these technologies will be developed are through individuals and teams who possess the requisite disparate knowledge base comprised of next generation sequencing, deep learning, and cellular and molecular biology (e.g. a full stack biotechnologist).

Another practical application of artificial intelligence to biotechnology can be found within the discipline of metabolic engineering. This area of study involves the development of novel synthetic pathways for the development of biomolecules through engineering the genomes of organisms particularly suited to produce a desired product. These organisms are then scaled up to produce quantities of the desired product either through intracellular or extracellular pathways. Metabolic engineering is the precursor to developing a competent cell line necessary for biopro-cessing, whereby a competent cell line must first be developed and scaled-up before a novel product can be successfully manufactured to produce a biomolecule at functional quantities.

Metabolic engineers around the world have the potential to develop harmful agents and novel synthetic pathways to derive pathogenic organisms or dangerous molecules and it is important to consider from a biosecurity perspective that this skill set his highly in demand across multiple market sectors including but not limited to oil and gas, pharmaceuticals, chemicals, and life sciences.

5 Moore's Law and the Convergence of Electronics and Biotechnology

Moore's Law is essentially the concept that the number of transistors in an integrated circuit design will double nearly every two years [21]. In the early 1990s, Tom Knight, an electrical engineer-turned-biochemist predicted that there is a limit to the ability to scale down silicone transistors and that Moore's Law would come to a logical end and biochemical systems would pick up where silicone leaves off [20]. In 2021, Tom Knight enjoys seeing the fruits of his prediction as one of the co-founders of the fastest growing synthetic biology company in the world, Ginkgo Bioworks. Knight's foresight was unique in that he possessed the expertise in electrical engineering to successfully predict its limitations. As a result, he started working in biochemistry, which routinely and naturally performs incredibly advanced processes at a scale smaller than the limits anticipated for integrated circuit design. The effects of Moore's law coming to an end is starting to be fully realized as the largest and most successful silicone chip manufacturer (Intel) in the world struggles to develop new chips at sizes below 10 nm [7].

Biotechnology not only has the capability to pick up where Moore's law will plateau, it is already doing so in the forms of novel biosensor systems, cell-free reactions, DNA-storage technologies, and emerging neurotechnology applications.

5.1 Novel (Bio)Sensor Systems

Biosensors combine the specificity and sensitivity of biological systems with the sensing capabilities often utilized with microprocessors [27]. Compared to electronic sensors, biosensors are significantly more sensitive and specific. They can monitor the environment, sense components of human and animal health, and detect a myriad of ligands dependent on their design. The most widely used biosensor is the glucose test trip, used to monitor a patient's blood sugar. Other applications of biosensors include sensing of airborne bacteria, sensing of water quality, detecting levels of toxic substances during bioremediation and more. Biosensors consist of biological and often also chemical or electronic components used for transduction of the sensor signal into usable information. In recent years, the convergence of materials chemistry, polymer chemistry, molecular biology, and electrical engineering have resulted in consumer-focused protein-based biosensors for transdermal alcohol detection systems [25], to aptamer-based biosensors for optical and mass-sensitive analytical techniques [22].

Further, cell-free biosensor systems are being developed for novel applications, which are rapidly reducing costs associated with developing whole organism biosensors.

Biosensors will continue to be used for national security applications, perhaps as early warning detection systems for a biological or chemical weapons attack. Of course, biological sensors are only useful if they are distributed in the appropriate locations for sensing, so distributed biosensor networks will likely emerge as a key component to biosecurity risk mitigation and threat detection applications.

5.2 DNA Storage Technologies

Another field currently converging with biosecurity is cybersecurity. The emerging discipline of Cyberbiosecurity is being explored for potential applications in many of the areas and markets discussed within this chapter [17]. A critical aspect of cybersecurity is the notion of data availability and one risk to data availability is the destruction or loss of data. Enter another biotechnological advancement emerging to address that concern. With a shelf life of more than 500 years, it is now possible to write, store, and decode high density information stored in DNA [2].

In 2012, George Church and his colleagues gained media buzz after encoding a draft of his 50,000-word book into binary code and converted it to a DNA sequence. Zeros were coded as C or A and ones were coded as T or G. This sequence was subsequently written on to a microchip in oligonucleotide fragments with an inkjet DNA printer. The team later decoded the fragments with PCR and successfully translated the text back to binary. This demonstration held around 650 kB of data and allowed the team to hypothesize that they could store 700 terabytes per cubic millimeter [6]. Although this experiment was performed first in 2012 as a proof-of-concept, it has been repeated successfully more recently as well [18]. This storage capacity-to-size ratio is considerable, compared to a modern sixteen terabyte hard drive in 2019, which weighs nearly 5 lb and has a cubic volume of nearly 3 cubic liters.

In addition to its small size, its extended half-life (compare to traditional archival tapes) is especially attractive to companies like Microsoft, which have an interest long term storage and the adoption next generation biological technologies for the longevity of its business [26].

5.3   Neurotechnology

Neurotechnology is another particularly relevant and emerging field in biotechnology. Neurological interfaces are being developed around the world for research, entertainment, and military purposes. This nascent field has particularly piqued the interest of the DARPA, which has developed a program to support the development of new interfaces for interacting with machines [23].

Emerging neurotechnologies can be classified as invasive, minutely invasive, or entirely non-invasive (e.g. transcranial direct current stimulation) and pose various biosecurity, privacy, and medical risks to users depending on their mode of application and intended purpose.

The biosecurity considerations associated with emerging neurotechnologies are particularly relevant at a time when cybersecurity flaws resulting in privacy breaches and identity theft are in the media cycle ad nauseum. Consumers of today's world may simply be concerned about the risks associated with DNA sequencing companies stealing their information [3]; imagine the risks, threats, and vulnerabilities associated with a world where augmented reality supported by neurological interfaces offers seamless, usable, enjoyable, and friendly user experiences.

6   Summary

Biotechnology is fast emerging as a successor to technology. In no other field is there such a tremendous convergence of different systems and disciplines for the development of new and novel technologies. As a result of this convergence and rapid development of new technologies, new risk and threat domains will continue to emerge for consideration and mitigation by biosecurity professionals. In the immediate future, artificial intelligence and bioinformatics will continue to be the tool of choice to safeguard the world from existing and novel biotechnological threats. As time progresses, it will likely be biosensors, cell-free systems, novel organisms, and advancements in neurotechnology that will safeguard our future from threats created by the same technologies. History has shown that the best way to safeguard against biological threat is through biological mitigation strategies and this will continue to hold true long after Moore's Law has reached its long-anticipated plateau.

References

1. Bonime W (2018) Bolt threads launches its first Mylo? leather product with a stylish tote bag. https://www.forbes.com/sites/westernbonime/2018/09/08/bolt-threads-launches-its-first-retail-product-with-the-mylo-driver-bag/
2. Bornholt J, Lopez R, Carmean DM, Ceze L, Seelig G, Strauss K (2016) A DNA-based archival storage system. ACM SIGOPS Oper Syst Rev 50(2):637–649
3. Brown K (2017). 23andMe is selling your data, but not how you think. https://gizmodo.com/23andme-is-selling-your-data-but-not-how-you-think-1794340474
4. Cello J, Paul AV, Wimmer E (2002) Chemical synthesis of poliovirus cDNA: generation of infectious virus in the absence of natural template. Science 297(5583):1016–1018
5. Chen F, Tillberg PW, Boyden ES (2015) Expansion microscopy. Science 1260088
6. Church GM, Gao Y, Kosuri S (2012) Next-generation digital information storage in DNA. Science 1226355
7. Cranz A (2018) Why the heck is intel struggling to make smaller, faster cpus? https://gizmodo.com/why-the-heck-is-intel-struggling-to-make-smaller-faste-1825597289
8. Duffy DC, McDonald JC, Schueller OJ, Whitesides GM (1998) Rapid prototyping of microfluidic systems in poly (dimethylsiloxane). Anal Chem 70(23):4974–4984
9. Fong C (2018) Open insulin, battle of the DIY diabetic. http://www.makery.info/en/2018/04/24/open-insulin-la-bataille-du-diabete-diy/
10. Fun GCAT (n.d.). https://www.iarpa.gov/index.php/research-programs/fun-gcat
11. Ganiukova K (2018) What is a full stack developer in 2018 and how to become one? https://hackernoon.com/what-is-a-full-stack-developer-in-2018-and-how-to-become-one-ca82e8906c87
12. Johnson B (2018) Re-writing nature's recipe book: how Ginkgo Bioworks is poised to upset almost every industry you? https://medium.com/future-literacy/re-writing-natures-recipe-book-how-ginkgo-bioworks-is-poised-to-upset-almost-every-industry-you-4e516a78bebb
13. Kitson PJ, Marie G, Francoia JP, Zalesskiy SS, Sigerson RC, Mathieson JS, Cronin L (2018) Digitization of multistep organic synthesis in reactionware for on-demand pharmaceuticals. Science 359(6373):314–319
14. Leitão P (2009) Agent-based distributed manufacturing control: a state-of-the-art survey. Eng Appl Artif Intell 22(7):979–991
15. Martin JH, Yahata BD, Hundley JM, Mayer JA, Schaedler TA, Pollock TM (2017) 3D printing of high-strength aluminium alloys. Nature 549(7672):365
16. Mueller B (2012) Additive manufacturing technologies–rapid prototyping to direct digital manufacturing. Assem Autom 32(2)
17. Murch RS, So WK, Buchholz WG, Raman S, Peccoud J (2018) Cyberbiosecurity: an emerging new discipline to help safeguard the bioeconomy. Front Bioeng Biotechnol 6:39
18. Offord C (2017) Making DNA data storage a reality. https://www.the-scientist.com/cover-story/making-dna-data-storage-a-reality-30218
19. Pearce JM (2012) Building research equipment with free, open-source hardware. Science 337(6100):1303–1304
20. Rogers B (2017) Tom Knight's Ginkgo Bioworks seeks to reinvent Moore's law through biochemistry. https://www.forbes.com/sites/brucerogers/2017/01/13/tom-knights-ginkgo-bioworks-seeks-to-reinvent-moores-law-through-biochemistry/#4af81937a938
21. Schaller RR (1997) Moore's law: past, present and future. IEEE Spectr 34(6):52–59
22. Song S, Wang L, Li J, Fan C, Zhao J (2008) Aptamer-based biosensors. TrAC Trends Anal Chem 27(2):108–117

23. Strickland E (2018) DARPA wants brain interfaces for able-bodied Warfighters. https://spe ctrum.ieee.org/the-human-os/biomedical/bionics/darpa-wants-brain-interfaces-for-able-bod ied-warfighters

24. Sullivan D, Igoe T (2004) Physical computing: sensing and controlling the physical world with computers. Thomson, Boston

25. Tanksalvala S (2018) Students develop wristband that glows when wearers are too drunk to drive—CU denver today. https://www.cudenvertoday.org/students-develop-wristband-glows-wearers-drunk-drive/

26. Tung L (2018) Microsoft's DNA storage breakthrough could pave way for exabyte drives| ZDNet. https://www.zdnet.com/article/microsofts-dna-storage-breakthrough-could-pave-way-for-exabyte-drives/

27. Turner A, Karube I, Wilson GS (1987) Biosensors: fundamentals and applications. Oxford University Press

28. Wegrzyn R (n.d.) Safe genes. https://www.darpa.mil/program/safe-genes

Technological Advances that Test the Dual-Use Research of Concern Model

Kavita M. Berger

Abstract Although malicious use of beneficial science and technology capabilities has elicited concern for many decades, the scientific and security communities have struggled with the review and oversight of dual use life sciences research. During the past twenty years, changes in the biotechnology landscape have increased concern about its exploitation. These changes are: (a) the rapidity with which biotechnologies are emerging, developing, and being applied; (b) the diversity of stakeholders involved in funding, conducting research and development activities, and using the new biotechnologies; and (c) some nonstate actors have expressed interest in using biology to cause harm and a few states have demonstrated use of biotechnology advances for harmful purposes. The multidisciplinary and global nature of life sciences and biotechnology, and diversity of private and government funders of research challenge recently developed approaches for identifying and assessing the dual use potential of biotechnology. Furthermore, development and application of biotechnology are driven by diverse societal needs including medicine, agriculture, environmental remediation, and defense. Entities invest in and leverage newly developed capabilities, such as precise genome editing or data analytics, to address critical needs in their areas. Within this context, this chapter explores challenges to the discourse on dual use implications of life-science research because of technology developments and provides a practical approach for assessing the potential dual use risks of emerging biotechnologies.

1 New Technologies and the Dual-Use Dilemma

Ten years after publication of *Biotechnology Research in an Age of Terrorism*, which introduced the concept of malicious exploitation of legitimate life science research

K. M. Berger (✉)
Gryphon Scientific, LLC, Takoma Park, USA
e-mail: kberger@nas.edu

The National Academies of Sciences, Engineering, and Medicine, Washington, DC, USA

© Springer Nature Switzerland AG 2021
R. N. Burnette (ed.), *Applied Biosecurity: Global Health, Biodefense, and Developing Technologies*, Advanced Sciences and Technologies for Security Applications,
https://doi.org/10.1007/978-3-030-69464-7_8

(termed 'dual use dilemma'), the international scientific community seemed to have its first true test of dual use life sciences research. In 2011, two highly respected, senior scientists—one in the U.S. and the other in the Netherlands—became the center of attention because of their research, which involved identification of specific mutations in the hemagglutinin (H) 5 protein of influenza virus that enabled transmission of H5 viruses in mammals. The studies were flagged by *Nature* and *Science*, the two journals to which they submitted manuscripts for publication, as having dual use information and the editors consulted the U.S. National Science Advisory Board for Biosecurity (NSABB) for further review and guidance. Initial review by the NSABB focused on the potential dual use risks and resulted in a split decision for the paper from the U.S. laboratory but unanimous concern for the paper from the Dutch laboratory. Their recommendation was to redact information that they felt could be directly misapplied prior to publication. Their recommendation raised several policy questions, among them whether the information should be export controlled and whether the U.S. policy to allow for open communication of fundamental research should be upheld.

Thirty years before these discussions on the H5 influenza mutations, the U.S. National Research Council published a report on *Scientific Communication and National Security*, which described the national security risks of mathematics and physics using the example of cryptography research and encryption. Then President Reagan issued National Security Decision Directive-189 (NSDD-189), which stated that the U.S. policy is to not classify fundamental research that is intended to be communicated broadly with the scientific community. Although every Administration after Reagan's had upheld the policy, the NSABB recommendation to redact certain information in the H5 influenza papers called into question the relevance of NSDD-189. However, the Obama Administration chose to uphold the policy. Still, questions about export control of information recommended-to-be-redacted remained and influenced subsequent international dialogue on the H5 influenza papers. The Netherlands took a different approach; they required the Dutch laboratory to obtain an export license for verbal and written communication of the research [29].

Following initial reviews by NSABB, the World Health Organization (WHO) convened meetings of international and public health stakeholders to explore the benefits of the influenza research as compared to the dual use risks. One benefit of the research lauded by the U.S. and Dutch scientists and the U.S. government sponsors of the research was in enhancing detection of naturally occurring H5 avian influenza viruses that could result in widespread transmission of the virus in humans. The researchers and sponsors argued that the mutations identified could be used by public health officials and influenza surveillance experts to signal the emergence of a human transmissible H5 influenza virus. The director of the U.S. National Institute of Allergy and Infectious Diseases, the scientists, and the editors of the U.S.-based journal *Science* had to obtain export control licenses to attend the international WHO meetings and discuss key aspects of the papers. Ultimately, the WHO and its public health stakeholders stated that the benefits of such research outweigh the potential risks of the research. After considering the WHO's conclusions, the NSABB recommended the papers be published, but with modification of the text in one of the articles. Around the same time, the White House released the *United States Government Policy for Oversight of Life Sciences Dual Use Research of Concern*, which directed federal agencies funding life science research to develop oversight and review processes for assessing and mitigating dual use potential of research involving any of 15 listed pathogens and any of 7 traits or experiments of concern. The companion policy, the *United States Government Policy for Institutional Oversight of Life Sciences Dual Use Research of Concern*, was directed towards research institutions and was released in 2014.

The H5 influenza papers emerged in the U.S. policy discourse again in 2014, approximately one year after they had been published. This time, two high-profile events—both accidental transfers of infectious viruses from U.S. laboratories—provided an opportunity for the U.S. government to re-examine dual use life sciences research involving pathogens that may cause a local or global epidemic if accidentally released from the laboratory. This research, dubbed "gain-of-function" (GOF), became the focus of the U.S. government, the NSABB, scientists, bioethicists, and the international health security community for more than two years. In the fall of 2014, the White House Office of Science and Technology Policy issued a policy to halt future funding of GOF research involving three respiratory pathogens (H5 avian influenza virus, Severe Acute Respiratory Syndrome (SARS) coronavirus, and Middle East Respiratory Syndrome (MERS) coronavirus) until the risks and benefits of such research could be assessed adequately.

This policy set in motion several efforts, a few of which informed the eventual U.S. government policy, *Recommended Policy Guidance for Departmental Development of Review Mechanisms for Potential Pandemic Pathogen Care and Oversight* 2016. Shortly after the research pause was issued, the U.S. National Institutes of Health (NIH) extended the pause to include currently funded research that met the criteria in the GOF policy document. The NIH initiated three activities that converged to inform the NSABB about the risks, benefits, ethical considerations, and stakeholder views associated with GOF research. The NSABB remained engaged during the entire process, allowing them to make specific policy recommendations about whether and under what conditions should GOF research with respiratory pathogens be conducted. These recommendations, which were issued in 2016, informed the development of the U.S. policy. At that time, the fourteen U.S. agencies that fund life science research were tasked with establishing their own processes for implementing the federal policy.

The research that led to this fervor of policy activity involved more basic experimental procedures. Specifically, both involved placing ferrets infected with H5 influenza virus in close proximity and isolating the variants demonstrating sustained transmission between ferrets. The U.S. laboratory evaluated the biosafety risks in advance of research initiation and created a laboratory-made virus that could be controlled by Tamiflu (the recommended antiviral drug for treating influenza infections). The Dutch laboratory used the code of conduct created by the Netherlands Academy of Sciences to evaluate risks of their research and obtained a containment box in which the research with the natural variant of the virus was conducted.

No advanced biotechnologies—such as 3D bioprinting, synthetic genomics, genome editing, or data analytics—were used in the H5 influenza virus studies, raising questions about whether and how to understand the potential risks of research involving newer technologies. According to current U.S. dual use life sciences policies, studies involving at least one of the 15 listed pathogens would be subjected to review and oversight. However, studies that do not involve any of the listed pathogens fall outside the policy's purview. In practice, U.S. funding agencies and research institutions are obligated to review and oversee only research that fits within the defined criteria of the policies. Furthermore, the policy focuses on research involving a defined set of pathogens supports this limited applicability of the dual use policies to the rapidly growing life science research enterprise, leaving many in the security community wondering how to evaluate the security implications of advances in biotechnology, convergence of biology with other scientific disciplines (e.g. computer science, mathematics, physics, and engineering), and new forms of research and services (e.g., automated laboratories, 3D printing of biomaterials, and custom-made viral reagents).

The desire to understand the security implications of advancing scientific information and technologies is not new. More than seventy years ago, policymakers and physicists were consumed with advances in theoretical physics and its applications to atomic energy and weapons. On the one hand, investments in defense provided the capital to drive researchers of the Manhattan Project to understand and ultimately control the creation of atomic energy, both necessary for the development of atomic weapons. With every scientific success, funding and support increased despite growing concern among scientists involved with the effort about the potential devastation that these weapons could cause. As policymakers sought to leverage scientific advances, several, but not all, leading scientists engaged in discourse with the scientific and policy communities about the ethics of developing weapons that could destroy entire villages, cities, or populations. They published editorials in *Science* and other prominent journals to convey their concerns, raise awareness of potential risks, and prevent proliferation of destructive weapons. Soon after the U.S. dropped the nuclear bombs on Japan, effectively ending World War II, governments negotiated the Nuclear Nonproliferation Treaty to prevent further development of nuclear weapons by countries that did not have such capabilities. Although this chapter does not focus on nuclear security issues, the efforts involved in understanding and communicating the broader implications of research and development in nuclear physics highlights the long history of evaluating emerging research and technologies.

Every time a scientific field begins to generate new information and technologies, concerns about the broader social implications follow. The field of nuclear physics is just one example. Other examples include the emergence of genetic engineering in late 1960s and early 1970s, when scientists developed the capability of making or changing an organism's DNA sequence using new tools and methodologies for cutting and pasting DNA; cryptography in the 1980s, which raised significant concern about revealing encryption codes; and new biotechnologies in the late 1990s through today, which elicited concern about simplifying viral or bacterial modification or creation.

The drivers for each of these examples is different, but the developments nevertheless resulted in active engagement by the scientific and security communities to discuss the potential security implications. In the 1970s, a small group of scientists had been studying a specific cancer-causing animal virus using new tools for cutting and pasting DNA (i.e., DNA splicing), the results of which elicited significant concern among the scientists. Fearing the development of regulations, the scientists called for a moratorium on recombinant DNA until they could develop a framework for conducting genetic engineering research safely and securely. Although much of the discussion focused on safety, the attendees of the conference were aware of the potential security concerns, having included some level of discussion on these topics. This meeting, referred to the Asilomar Conference on Recombinant DNA, resulted in a set of rules for self-governance, which eventually became the NIH recombinant DNA guidelines and informed parts of the Biosafety in Microbiological and Biomedical Laboratories (BMBL) manual. Shortly after the Asilomar Conference, concerns about possible dangers were raised among policymakers and the public, resulting in articles about the issue and introduction of legislation to minimize harms [6]. However, the development of the NIH guidelines subverted those efforts.

In 1977, the Institute of Electrical and Electronics Engineers (IEEE) held a symposium on cryptography during which scientists discussed new encryption methodologies based on computer science concepts. An employee of the National Security Agency attempted to stop the symposium by stating that publication of this information would be a violation of export control regulations and the scientists could be prosecuted under the Arms Control Act of 1976. This was one of several events that occurred between 1977 and 1982 in the field of cryptography and encryption research. The NSA was concerned that open publication of encryption techniques could improve foreign government encryption capabilities making code-breaking difficult, help foreign governments identify vulnerabilities in their encryption methods, and enable foreign governments to identify and exploit any vulnerabilities in U.S. methods. Together, these events and concerns changed the handling of fundamental research that is thought to produce national security risks. The major changes were: (a) the requirement that a government employee, contractor, licensee, or grantee of an agency that does not have classification authority must protect information that they believe should be classified; (b) the establishment of a process through which the National Science Foundation, a major funder of cryptography research, and the NSA could share proposals that may raise national security concern and consider recommendations for not funding such research if a sufficient rationale was provided; and (c) the creation of a system whereby cryptography researchers voluntary would submit manuscripts to NSA for review and to journals at the same time. As these changes were happening, the National Research Council established a committee to examine scientific communication of fundamental research that may have potential national security implications. The National Research Council's committee efforts led to the issuance of National Security Decision Directive 189 (NSDD-189), which states that fundamental research should not be classified if it normally would be published in open literature. In addition, this strategy defined fundamental research, distinguishing it from proprietary or classified research.

The next significant challenge came in the 1990s when the world realized that the Soviet Union had supported a massive and advanced offensive bioweapons program, saw a Japanese cult try to develop and use a biological agent as a weapon in Tokyo, and was unable to prosecute a white supremacist seeking to obtain plague bacteria illegally. These events, together with biotechnology becoming a dominant field worldwide, led members of the scientific and security community to examine the potential security risks of life science research. To examine these concerns, the U.S. National Research Council established a committee to evaluate the potential exploitation of biotechnology by individuals or groups with malicious intent. During the committee's deliberations, the U.S. was attacked by a terrorist organization followed by a bioterrorism incident involving anthrax spores in letters sent through the mail, publication of the first chemical synthesis of live poliovirus by life scientists, modification vaccinia virus (a version of the original vaccine for smallpox) by inserting a gene from smallpox by another research group, and creation of a vaccine-resistant mousepox virus by Australian scientists. These events influenced the committee's recommendations, which introduced the world to the dual use dilemma in the life sciences. Unlike past examples where dual use referred to civilian and military or intelligence applications, the dual use dilemma was defined as legitimate research that could be exploited by malicious actors (primarily, non-state actors) to cause harm to others. In 2004, just four years after the 9/11 attacks, the U.S. National Research Council published the committee's findings in its report, titled *Biotechnology Research in an Age of Terrorism* [27]. The report identified specific types of experiments that could be misapplied by malicious actors, which became the foundation for U.S. policies on dual use life sciences research and called for the establishment of a national-level advisory board to consider harmful application of life sciences research. The visibility and timeliness of the report contributed to significant changes in biosecurity in the U.S. and internationally by broadening the dialogue from a sole focus on prevention of offensive biological weapons to prevention of malicious exploitation. This new conception involved a broader array of stakeholders and focused on responsible science and codes of conduct as potential prevention measures.

Around the same time, the American Society for Microbiology (ASM) convened a meeting on publication of research with dual use potential and included scientific leaders, biosecurity experts, and editors from premier journals, including *Science, Nature*, Cell, and the ASM journals [2]. The meeting served two purposes: (1) raise awareness about the possibility that scientific information could present risk in addition to benefit; and (2) identify a common set of principles for publication of scientific information about which journals and authors could agree. The attendees of the meeting co-authored a statement highlighting four key considerations in publication of life science research: (1) protect the integrity of the scientific process by publishing high-quality manuscripts that include sufficient detail to enable reproducibility of the experiments and analyses; (2) increase their capacity to identify and address concerns about safety and security that may be raised in published papers; (3) encourage the design of review processes that enable identification of safety and security risks of research; and (4) modify or not publish manuscripts that contain information that could present greater potential harm than benefit to society. This statement became the first of several activities focused on reducing the potential security risks of life science research.

Shortly after these efforts, the White House established the National Science Advisory Board for Biosecurity under the auspices of the U.S. National Institutes of Health. The original charter was modeled after the Recombinant DNA Advisory Committee and enabled the NSABB to review advances in life science research to identify potential biotechnologies that deserve greater oversight because they present security risks and to review and oversee specific research that may pose security risks. Throughout the next decade, the NSABB developed recommendations for criteria of dual use life sciences research, strategies for communicating and overseeing such research, and outreach approaches educating scientists about the dual use dilemma. In addition, the NSABB was asked on occasion to review specific research. These efforts were intended to provide the U.S. government and the scientific community with guidance on detecting and mitigating research with dual use potential.

As the committee deliberated, the rapid advances in and decreasing costs of DNA sequencing and synthesis became a primary driver for dramatic changes in the types of individuals conducting research with biological materials and the venues they used. Computer scientists and engineers began playing around with "biological parts" to determine whether biological systems could be put together in a predictable manner and based on known sequence and gene function. This effort led to the creation of the international Genetically Engineered Machine (iGEM) competition. Amateur biology became more popular as individuals with little or no life-science training began extracting DNA from common items using household chemicals. However, access to research materials proved challenging to the amateur biology community, which led to the establishment of community labs, the development of an industry to support the amateur biology community, and the creation of an affiliation group, Do-It-Yourself Biology (DIYBio). Whether working in community labs or their homes, amateur biologists began making simple genetic modifications in bacteria and yeast and sequencing DNA from food, residue on cups and other similar items, and the environment. At the same time, professional scientific organizations, such as the J. Craig Venter Institute, began sequencing microbes in the ocean to identify new species of living organisms and pushing the envelope for synthesis of microbes.

The methods for synthesizing genes and for chemically linking synthetic pieces of DNA continued to improve, raising significant concern among the security community about potential exploitation of these approaches by individuals with little scientific skill. This concern led the U.S. government to ask the NSABB to evaluate the potential that synthetic genomic techniques could be used to make controlled, highly restricted pathogens. The NSABB recommended that the U.S. National Academy of Sciences conduct a study to determine whether pathogen virulence could be predicted from genetic sequence, the U.S. government consider whether and how DNA synthesis orders could be monitored for the controlled pathogens, and the NIH include synthetic nucleic acids in the Recombinant DNA Guidelines. All three recommendations were implemented, resulting in: 1) a National Research Council report stating that current bioinformatics capabilities and scientific information are insufficient to predict function reliably and repeatedly from DNA sequence; 2) the Department of Health and Human Services Guidance for Providers of Synthetic Double-stranded DNA; and 3) addition of synthetic nucleic acids in the Recombinant DNA Guidelines. In addition, the Federal Bureau of Investigation (FBI) began reaching out to synthetic biology practitioners to raise awareness of security risks and advocate for vigilance within their community to safeguard their work from careless and/or malicious individuals seeking to exploit biology to harm others. The Defense Advanced Research Projects Agency (DARPA) established a research program to develop technical solutions against the spread and persistence of synthetic organisms in the environment. This effort is described later in this chapter.

2 Security Concerns About Emerging Biotechnology

In the early twenty-first century, discussions about security risks of emerging biotechnology mostly focused on synthetic genomics and synthetic biology. However, other organizations, such as the U.S. National Academy of Sciences and InterAcademy Partnership conducted workshops to discuss biotechnologies of possible relevance to the Biological and Toxins Weapons Convention. Still others, such as Margaret Kosal and the late Jonathan Tucker, published books on potential exploitation of emerging chemical and biological technologies. Although some of these efforts focused on modification or creation of existing or new pathogens, others included technologies such as genomics, nanotechnology, and gene delivery (e.g., gene therapy vectors) in their analyses. The results of these scholarly efforts remain to be seen as international and domestic policy efforts occur at a much slower pace than advancement and application of these and other biotechnologies.

Today, the security policy community (both within the U.S. and internationally) are focusing their attention on genome editing, especially the use of clustered regularly interspaced short palindromic repeats (CRISPR) based tools. CRISPR-based tools were developed in 2014 to cut DNA at precise locations, enabling repair or replacement of specific DNA sequences. This capability is most advantageous in modifying eukaryote genomes primarily because other genetic engineering approaches can be as, or perhaps more, efficient at modifying viral and bacterial genomes, and bacterial cells do not have the same ability to repair cut DNA as eukaryotic cells. However, the security community has expressed concern that CRISPR-based tools, and other genome editing tools, could be used to make never-before-seen pathogens or to make harmless pathogens harmful. Former Director of National Intelligence General James Clapper described gene editing as a weapon of mass destruction in his 2016 Congressional testimony on Worldwide Threats. Furthermore, the subsequent Director of National Intelligence, Mr. Coates, described gene editing as an emerging technology to monitor in his 2017 Congressional testimony of Worldwide

Threats. This high-level of concern prompted several governmental efforts evaluating the potential national security risks of genome editing, all informing policy deliberations on risks and potential regulatory action of genome editing.

Although most of these efforts focus on human health outcomes (illness and death) as the primary parameters for assessing national security risks, the intelligence community also considers economic and commercial, military, social, and political risks. These risks open the door to considering new facets of synthetic biology and genome editing and consequences of biotechnologies whose primary applications do not involve pathogens. For example, in 2014 the FBI, American Association for the Advancement of Science (AAAS), and United Nations Interregional Crime and Justice Research Institute (UNICRI) undertook a study to evaluate the potential risks and benefits of big data in the life sciences. This study was driven by the rapid advances in genomic medicine and learning health care system, and seemingly daily news of cyberattacks of health care databases. The FBI, the project sponsor, was interested in understanding the types of risks enabled by big data applications, the benefits of such technologies for enhancing national security (e.g., pathogen surveillance and medical countermeasure research, development, and testing), and the current governance structure that minimized risk and maximized leveraging of benefits. This study led to the development of a qualitative evaluation of risk and benefit, development of illustrative case studies, and worksheets for security experts and scientists to use when considering risk and benefit.

3 The Biotechnology Enterprise

The global enterprise for biotechnology includes knowledge, products, and services developed by academia, industries, citizen scientists, entrepreneurs, and governments, resulting in a research, development, and commercial market well above tens-of-billions of dollars. The funding landscape similarly is diverse with the venture capital community resulting in new models for promoting innovation among entrepreneurial individuals interested in leveraging biology to address a particular need, and with crowdsource funding websites such as Kickstarter, enabling increasingly sophisticated research activities by individuals. Furthermore, the biotechnology enterprise contributes to the health, agriculture, energy, environmental, science, and defense sectors. The multidisciplinary nature of modern biotechnology drives advancement in the basic and applied sciences, enabling new approaches for addressing health care and public health needs, enhancing production of agricultural commodities, pushing scientific research and analysis, preserving biodiversity and ecological health, and protecting military personnel and materiel.

3.1   Contributions of Biotechnology to National Security

Biotechnologies provide critical capabilities in the defense against natural or man-made threats. In health, new scientific discoveries and platforms contribute to the development of vaccines and medicines against pathogens, such as influenza virus, anthrax, and SARS-CoV-2. Researchers' ability to understand the relationship between pathogens and their hosts provide the information needed to create targeted vaccines and therapeutics. Similarly, scientists have begun applying synthetic biology tools and concepts to vaccine development, hoping to reduce the time involved in creating a vaccine against an emerging infectious disease and enabling rapid response efforts. More recently, investments in additive biomanufacturing have provided new opportunities to print human tissues and organs, which in some laboratories are being used to study the effects of pharmaceutical candidates before testing in animals or people. In addition, advances in knowledge gained from neuroscience and behavioral sciences provides new opportunities to understand and identify clinical interventions to prevent or treat various psychological disorders, such as post-traumatic stress disorder after return from military conflict.

Beyond health efforts, governments are looking to the life sciences and biotechnology to assist in early warning of infectious disease events, forensics, decontamination of materials, and many other uses. Efforts to gain advanced warning of infectious disease events have leveraged new data analysis technologies (e.g., artificial intelligence and machine learning, natural language processing, image analysis, and text analysis) to evaluate a variety of data, including environmental, ecological, genomic, physiological, and other data. Efforts to promote longevity of materials and equipment have leveraged advances in synthetic biology. Efforts to enhance microbial forensics and attribution leveraged advances in microbiology and improved sample and data sharing procedures between public health and law enforcement communities. Enabling this research and data sharing was a priority for the U.S. government and international organizations to seek solutions for capability gaps in preventing, detecting, and responding to natural or man-made infectious disease events of international concern.

One of the fourteen U.S. government agencies that fund life science research is the Defense Advanced Research Projects Agency (DARPA), which promotes innovation in biotechnology to address national security needs, specifically focusing on the U.S. military [7]. These advances may have broader application to other sectors such as health, for which DARPA has a dedicated office for transfer of promising technologies developed from their investments. In 2014, DARPA established the Biological Technologies Office to explore research at the "intersection between biology and the physical sciences" and to "harness the power of biological systems." This office explicitly drives multidisciplinary science to develop more effective means of protecting the warfighter from exposure to biological and chemical hazards, improving diagnostics and field detection of biological agents, controlling or enhancing robustness of biological systems, and developing more effective medicines and prosthetics. DARPA's The Chronicle of Lineage Indicative of Origins (CLIO) and SafeGenes

[37, 8] programs are investing in basic and applied research to engineering technical safeguards for improving safety of synthetic biology and gene editing, respectively [9].

The DARPA CLIO program sought to develop long-lasting control elements that could prevent intentional harmful use of genetic engineering and acquisition of proprietary variants of organisms. This program supported research groups throughout the U.S. to develop various elements, such as DNA sequences that prevent survival of engineered organisms in the wild. This program responded to security community concerns about the application of synthetic biology to create unnatural or uncontrollable dangerous organisms. The knowledge and tools gained from these studies could be applied to environmental efforts to control laboratory-derived organisms and forensic efforts to assist with investigation of biological events. In addition, this program would allow funded researchers to demonstrate their actions towards ensuring safe and compliant research practices.

The DARPA SafeGenes program was launched in 2016 to develop technical approaches for preventing accidents and intentional harmful use of genome editing tools. This program is supporting research groups to develop approaches and technical solutions for controlling genome editing activity, including gene drive-mediated editing. In pursuit of this goal, research teams will increase basic research knowledge about how genome editing occurs, and which combinations of engineering controls reduce the greatest accidental or intentional risk. This program responds to security community concerns about the use of genome editing tools to create harmful organisms and environmental community concerns about uncontrolled release of edited organisms in the wild.

The benefits of biotechnology to national security apply to a range of activities from early warning of potential biological incidents to public health response in natural or man-made events.

3.2 Security Policy Considerations About Emerging Biotechnology

Advances and application of biotechnology and the changing face of stakeholders, both developers and users, presents various challenges for the scientific, security, and policy-making communities to assess the potential risks and reap the potential benefits. Ultimately, the global perception of security risk, which continues to focus on manipulating pathogens, leads many in the security community to focus on those technologies that can be used to modify and deliver pathogens or that they *perceive* could be used to modify or create pathogens. In contrast, organizations with specific needs for preparing for, detecting, and responding to biological threats often survey a wide variety of biotechnologies to determine which are suitable for their specific needs. The difference in scope of risk and benefit leads to a mismatch in review and governance of biotechnologies, which on occasion, has resulted in previously unanticipated security risks of beneficial biotechnologies. The remainder of this chapter will focus on the potential security risks of biotechnologies, paying close attention to the risk of exploitation of knowledge, skills, technologies, and commercial products and services that may result in consequences to human health, agriculture, economics, and/or social and political instability.

National security risks of biology historically have focused on deliberate weaponization of pathogens by nation states or non-state actors. This viewpoint largely stems from knowledge about the types of weapons developed and desired effects sought by the former offensive biological weapons programs. This view was reinforced in the early-1990s when the world discovered the massive covert biological weapons program of the former Soviet Union. Since then, the programs of a few other countries, including Iraq, Libya, and South Africa, were identified. During the 1980s and 1990s, incidents involving nonstate actors (i.e., Rajneeshee cult in 1984, Aum Shinrykio in the 1990s, and Larry Wayne Harris in 1995) also highlighted pathogens as a primary risk to national security. Programs to reduce the threat of biological weapons materials, knowledge, and skills were initiated to prevent individuals and groups with malicious intentions from acquiring certain pathogens. This basic infrastructure of pathogen-focused security changed dramatically after Autumn 2001, when the coincident attacks by Al Qaeda occurred and the letters containing anthrax spores were sent. Concerns about acquisition of harmful pathogens led to numerous legislative and regulatory efforts in the U.S. that resulted in the current iteration of the Federal Select Agent Program, and to several international engagement activities throughout the world to build local capability to prevent access to harmful pathogens and promote best biorisk prevention practices. Still, much of these efforts focused on pathogens considered to cause significant harm to human and animal health.

Throughout this time, members of the U.S. scientific and security communities began raising concern about the risk of harmful application of legitimate biotechnology, which the U.S. National Research Council termed "the dual use dilemma" in its 2003 report *Biotechnology Research in an Age of Terrorism*. Although the focus was on harmful application of pathogen research, this report (and subsequent policy activities) broadened the security discussion to include traits of concern and any scientific knowledge, skills, and technologies enabling the manipulation or creation of pathogens to obtain those traits. This initial effort, along with a parallel effort carried out by a group of concerned journal editors and authors, ultimately resulted in the development of U.S. policies on review and oversight of dual use research of concern (2012 and 2014), research involving certain variants of highly pathogenic avian influenza (2015), gain-of-function research with respiratory pathogens (2016), and research with pathogens of pandemic potential (2017). Together, these policies focus review and oversight guidance on certain types of research involving pathogens that could cause significant harm to public health and safety. In addition, they often describe explicitly the experimental approaches that may be used to achieve traits of concern in pathogens of concern, and they still are focused on preventing serious consequences to human and animal health. The U.S. government and qualifying research institutions are in the process of developing and implementing these policies. Ideally, these policies would be integrated into a single harmonized guidance by the U.S. government, similar to the Common Rule for human subjects protection. However, practically, each funding agency will create their own guidance for how to review and oversee research with dual use potential, with which the research community must comply. This means that as many as 14 agency-specific guidances could be developed for each of the aforementioned policies, which could result in a compliance challenge for institutions that receive funding from more than one of these agencies.

In addition to these efforts, experts from the U.S. and Europe have been engaging with scientists throughout the world to policies, review practices, and educational tools for training local scientists about dual use research of concern. The Biological Weapons Convention and the United Nations Interregional Crime and Justice Research Institute have provided international platforms for sharing information about these efforts. The primary challenges with wide-spread adoption of these tools are: (1) developing resources that are relevant to the local context, including institutionally and nationally relevant examples of dual use research and locally sustainable prevention measures; and (2) implementing programs in environments where other, higher priority concerns (e.g., active conflicts, limited research funding, lack of research positions) exist. Examples of educational programs that have addressed some of these issues are: a) the U.S. National Research Council International Faculty Development Project on Education about the Responsible Conduct of Science, which trained a selected group of scientists from Middle Eastern and North African (MENA) countries on adult learning methodologies and dual use research, encouraging them to develop and implement educational programs in their institutions [28]; b) the American Association for the Advancement of Science's practical training exercises on responsible science, which are based on high quality research published from MENA countries and intended for use with life scientists in the MENA region [24]; and c) Gryphon Scientific's training materials for practical implementation of biosafety and biosecurity measures [17].

As policies on restricted pathogens and dual use research of concern were being debated, concern about access to restricted pathogens or creation of extinct or unnatural pathogens arose, primarily from published reports on the chemical synthesis of poliovirus, de novo creation of a bacterial cell, and biological "circuits" that can produce chemicals and other small molecules. These efforts, which were described collectively as synthetic biology, elicited significant concern among security experts who feared that such capabilities could eliminate access barriers to harmful pathogens. Synthetic biology applies engineering principles (i.e., design, build, test, learn) to the manipulation of biological organisms, and involves computational design and robotics to conduct genetic engineering methods. This approach has enabled greater convergence of biology and chemistry, resulting in the creation of several organisms whose entire existence is to produce specific molecules for a variety of industries (e.g., cosmetic and chemical industries). It has spawned new companies that seek to create and/or use organisms as biofactories, automate laboratory functions, or supply custom-made products such as chemically synthesized genes or viruses. The emergence of synthetic biology also led to greater awareness of the growing amateur DIY biology community, which today includes community laboratories and individuals. A part of the DIY biology community conducts biological research for a specific purpose such as wanting to understand an inherited disorder within their families, whereas others in this community seek to play with biology, sometimes pushing the limits of ethics and safety. For example, one DIY biologist is known to experiment on himself.

The grouping of professional scientists, citizen scientists, and entrepreneurs under the synthetic biology umbrella has led many in the security community to believe that biology is becoming democratized and the access barriers to skills and technologies are decreasing, both of which have elicited concern about the ease with which individuals with little training and malicious intent could synthetize or make harmful pathogens. To address this concern, the U.S. government released voluntary screening guidance for providers of synthetic double-stranded DNA, which involves screening of the requested sequences and customers, and provides guidance about when to contact the Federal Bureau of Investigation if concerns arise. At the same time, the DNA synthesis industry joined together to develop industry guidelines for screening orders, which resembled the U.S. guidance. Screening of DNA sequences involves determining whether the requested sequences match the sequence of a restricted agent using bioinformatics tools. A primary difficulty in implementing these guidelines has been in distinguishing between sequences that could be used to make a pathogen more harmful and sequences that are considered common to harmful and harmless versions of the same pathogen type. In an attempt to address this challenge, three U.S. agencies have established research programs to develop databases linking function, in this case virulence, to specific genetic sequences. Although specific genes have been linked to virulence traits, creating a database of such information is technically challenging because genes behave differently in different genetic backgrounds and virulence traits of some pathogens are determined by the host, not the pathogen. However, if such a database were created successfully, the information contained within may present a security risk (i.e., a dual use information risk).

Advances in microbiology, virology, bioinformatics, and synthetic genomics represent a small portion of the overall biotechnology enterprise. The biotechnology landscape is vast, including omics and systems biology, personalized medicine and precision agriculture, computational analysis and modeling, advanced biomanufacturing, neuroscience, and nanobiotechnology to list a few. Current biosecurity governance does not overlap with biotechnologies that do not involve restricted pathogens or in some cases, data. The biosecurity-specific policies, the current export control regime, and ongoing policy debates focus on applications of biotechnology to the generation and manipulation of harmful pathogens, not biotechnologies and data that do not involve pathogens or toxins. However, as new capabilities to generate, analyze, and manipulate biological data and cells are developed, current conceptualizations of biosecurity likely will need to change. For example, the numerous breeches of health insurance and facility databases resulted in only patient health and genomic data being stolen. Security officials now are inquiring about why the data was stolen and for what the stolen data will be used. The purpose may be to gain competitive or economic advantage in the pharmaceutical or healthcare delivery sectors or to become the global leader in biotechnology and/or artificial intelligence disrupting global power dynamics of nation states.

4 Practical Applications of Dual-Use Considerations of Emerging Biotechnology

Current conceptualization of dual use life sciences research exclusively focuses on certain types of research involving pathogens. However, the idea that basic research could be used by adversaries to enhance military capabilities or to counter intelligence capabilities is not new. In the 1980s, concerns about the use of basic research on encryption by adversaries to decode protected intelligence information led the scientific community to evaluate the dual use potential of basic computer science and mathematics. The history and outcomes of these efforts are captured in the U.S. National Research Council Report *Scientific Communication and National Security*, which was published in 1982 [26]. Twenty-five years later, the scientific community and U.S. National Academy of Sciences were undertaking a similar evaluation for advances in biotechnology. Since 2000 when questions about dual use implication of biotechnology arose, scientists and security experts have engaged in various policy efforts to define dual use life sciences research and in educational activities to train scientists throughout the world about dual use issues. As highlighted in the previous sections, these efforts focused exclusively on pathogen research. Few efforts have focused on dual use implications of non-pathogen research, most of which have been scholarly rather than policy efforts. For example, the U.S. National Science Advisory Board for Biosecurity (NSABB) grappled with evaluating a paper from the field of organizational management that described the vulnerabilities of the U.S. milk supply to deliberate tampering. Some NSABB members thought this research did not conform to their conceptualization of dual use research while other did. Two other research groups—one from the University of Bradford and a second from Georgetown University and University of Pennsylvania—studied the national security implications of neuroscience. In 2012, the late Jonathan Tucker published a book on Innovation, Dual Use and Security, which explored dual use considerations for various chemical and biological technologies such as high throughput screening of chemicals, protein engineering, and bioregulators [34]. In 2014, the AAAS, FBI, and UNICRI published a report describing the benefits and security risks of big data in the life sciences.

4.1   Challenges of Applying Dual-Use Philosophies

Extrapolating these scholarly efforts to practical review of research has proven to be challenging for several reasons. First, little guidance is available to the research community about the types of consequences that are of greatest concern and the analytic methods for assessing the potential for harmful application of research. As described at length, policy and guidance is provided for certain types of pathogen research, which promotes "check-the-box" compliance. Second, the scientific community has expressed concern about assessing the potential for countries, malicious groups, and individuals to misuse research conducted for legitimate and beneficial reasons. Their concerns often stem from their identity as scientists attempting to learn about how the world works and how to prevent or treat disease. Many scientists explicitly state that, unlike the security community, their job is not to assess the potential harmful applications of research they believe to be legitimate or beneficial and the perpetrators of such harmful acts. Analogous sentiments have been echoed by members of the security community, but not policy experts who believe that scientists have a responsibility to review and minimize the risks (whatever they may be) of their research. Third, security policy experts have stated that scientists are not necessarily the most objective judges of the risks presented by their research. They have cited incentives for research progress and career advancement as biasing scientists' assessments. In addition, they have stated that scientists do not have the same level of appreciation for or awareness of security issues to be able to assess security risks adequately. At the same time, many scientists believe that the security community prioritizes possible risks over possible benefit, despite barriers that may exist in their realization of the benefits. The recent Risk and Benefit Assessment of Gain-of-Function Research describes the first ever framework for assessing benefits of research within the current regulatory, technical, and commercial context. Despite these challenges, one reality is true: if a harmful incident that involves or is enabled by biotechnology occurs, the consequences may be significantly worse than preventative assessment and mitigation of risk. Therefore, translating the scholarly assessments of security risks to practical applications can promote greater understanding of the potential risks of biotechnologies, and technical, organizational, and policy measures for mitigating risks.

The first step in applying scholarly approaches for assessing dual use risk of biotechnologies is to clarify the concerns about the potential harmful application(s) of the technologies. Relevant questions that can be asked to clarify the concerns include:

(1)   What harms could be caused by malicious use of the technology or data?
(2)   How severe can the consequences of the identified harms be?
(3)   Based on the current state of the technology and its applications, do the methods, information, and supporting tools enable harmful application of the technology?
(4)   What knowledge and technical barriers exist to prevent harmful applications of the technology?

(5)   Do measures exist to prevent or mitigate the harms?
(6)   Will these measures prevent the benefits from being realized or result in collateral damage?
(7)   Will prevention or mitigation measures stop scientific advancement and application of the biotechnology?
(8)   What types of adversaries could have the intent and means to use the technology to cause harm?

Using these guiding questions, policymakers, security experts, and scientists can begin to evaluate the potential security risks associated with biotechnologies and consider how these potential risks compare with the potential benefits. Although this approach does not enable quantitative assessments of risk, it provides an opportunity to clearly define the outcomes that present the greatest concerns. These outcomes may be adverse effects on a population's health, decreased or no production of major agricultural commodities, compromise of the public health or health care sectors, or other outcomes that may result in harmful social, economic, political, or health effects. In addition, analyzing biotechnologies using these questions enables a more flexible system for evaluating existing and future scientific advancements.

The second step is identifying mitigation strategies that address the security risks without adversely harming the advancement or beneficial application of biotechnologies. Once specific harms and vulnerabilities of biotechnologies have been identified, policymakers, security experts, and scientists can evaluate the potential policy and technical solutions for reducing the potential risks. In some cases, solutions already may exist to address specific vulnerabilities. In other cases, appropriate solutions may need to be developed or invested in, as may be the case for technical solutions to vulnerabilities. Still in some cases, the most effective solution may be ongoing monitoring and risk analysis of biotechnology advances and applications.

The third step is development and implementation of policy and/or technical solutions, evaluation of implementation efforts, and iterative improvement of solutions to address any gaps that may be realized during the development, implementation, and evaluation process. Iterative policy development at the federal and state government levels may be extremely challenging for laws and regulations. For example, the 2004 law that imposed penalties on possession, development, and storage of poxviruses that are 85% identical to variola virus (virus that causes smallpox) raised many questions about the scope of this law or any research involving poxviruses. Poxvirus researchers wondered whether they suddenly were in criminal violation of the law for possessing viruses in the same family as smallpox, including vaccinia virus (the original smallpox vaccine) and modified vaccinia virus Ankara (the current smallpox vaccine). The Department of Justice subsequently clarified this law, but questions about relevance might emerge in response to future scientific activities, such as the synthesis of the previously extinct horsepox virus [23]. At the other end of the spectrum is the NIH Guidance on Recombinant and Synthetic Nucleic Acids. This policy is required for all entities that receive federal funding for life sciences research, regardless of funding source of specific research projects, but is not a statute. During the past decade, the NIH office responsible for implementing this guidance has established working groups to discuss how emerging biotechnologies affect or are affected by the policy. As new scientific capabilities are developed, this policy is updated to ensure that it adequately covers new advances that raise specific *safety* (not necessarily security) concerns. However, this policy solution, which was developed in the 1970s by members of the scientific community and formalized by the NIH in the 1980s to address safety concerns associated with genetic engineering, has demonstrated its relevance to the changing life science research landscape for more than 40 years.

The following examples illustrate how this multi-step approach can be applied to new advances and applications of biotechnologies.

4.2 Synthetic Biology and Synthetic Genomics

In the early 2000s, the decreasing cost of DNA synthesis technology led to research that involved chemical synthesis of small viruses, a capability referred to as synthetic genomics. During the past 15 years, synthetic genomics was used to synthesize large microbes, including bacteria. Scientists throughout the world are pushing the envelope in DNA synthesis of organisms, including the creation of minimal genomes that support bacterial growth and recreation of large DNA and RNA viruses. At the same time, engineers, and later life scientists, began designing and assembling organisms to perform specific functions. These synthetic biology efforts have led to new industries, biotechnology products, and practitioners of biology (e.g., engineers, computer scientists, and citizen scientists). Synthetic biology and synthetic genomics have elicited many concerns by the security community about new acquisition pathways for restricted pathogens (i.e., via synthesis of the pathogen genome and "booting" up of a live organism). Furthermore, members of the environmental community have expressed concerns about accidental release of unnatural organisms and disruption of biodiversity. More recently, experiments describing the development of synthetic organisms that produce hazardous chemicals or illicit drugs have elicited significant concern about health consequences or criminal activity, respectively.

The severity of consequences ranges from low for creation of organisms that cannot persist in the environment, people, or other animals, to high for creation of organisms that could cause severe human health consequences or organisms that could produce illicit products that could breed lawlessness or criminal activity. Effective risk mitigation strategies for more severe harms involve a combination of physical security measures that prevent unauthorized access to higher-risk organisms, methods, and information; and institutional review and oversight of research to reduce the potential security risks of irresponsible communication and sharing of methods for creating high-risk organisms. However, these measures likely could counteract the scientific culture of research communication to promote scientific progress and contribution and enable researchers to demonstrate their knowledge to funders and tenure review committees. In addition, these measures could affect reproducibility initiatives adversely, inadvertently trigger export control of information even if the information is self-censored, and prevent transparency of high-risk research, breeding concerns about secret research programs. Therefore, the negative consequences of withholding information may be worse than communicating information with dual use potential. Active engagement among scientists, biosafety officials, and biosecurity experts about potential security risks of biotechnology and feasible (and not damaging) risk mitigation measures can help advance collective understanding of dual use assessments and useful mitigation measures.

The level of investment in synthetic biology research already has led to published methods and innovative products. Despite these advances, several technical challenges exist that prevent just anyone from conducting the more advanced research. These challenges include the relative unpredictability of associating specific genetic sequences with specific traits in different organisms, differences in gene expression in different organisms and plasmids, and the continuous need to troubleshoot experimental problems. Given these technical considerations, individuals must have the knowledge and skills to design, conduct, and troubleshoot synthetic biology and genomics experiments.

4.3 Genomics and Precise Genome Editing

Although genetic engineering has been developing since the late-1960s, the capability to edit a cell's genome precisely has exploded within the past six years. Protein-based tools, such as meganucleases, zinc finger nucleases (ZFNs), and transcription activator-like effector nucleases (TALENs), were developed initially. These proteins have two relevant domains: a DNA-binding section and a DNA-cutting enzyme. Although these proteins act differently to recognize and cut DNA, they all identify long sequences in the DNA and cut at another sequence that is a specific distance away from the DNA-binding site. Tailoring ZFNs and TALENs to different DNA sequences is possible, but difficult because modifying the DNA-recognition sequences in the proteins have to be made in the protein's DNA first, which is a process that can be time-consuming and biochemically challenging. However, within the past six years, scientists have created a new set of tools based on clustered regularly interspaced short palindromic repeats (CRISPR). These tools involve a two-component system with an RNA molecule that recognizes and binds DNA and a protein enzyme that cuts DNA. Initially, CRISPR-based tools were made with the Cas9 DNA-cutting protein. Today, several CRISPR-based tools exists, each with different DNA-cutting or modifying proteins. Some of these tools have more precise DNA-cutting enzymes, such as Cpf1, while others involve modifications of Cas9 that enable gene silencing, gene activation, inducible gene expression, and modification of the DNA structure. Within the last few years, scientists have created a version of the CRISPR-genome editing tool that edits RNA, an intermediary molecule between the DNA genome and proteins. The promise of these tools lies in their accuracy for identifying and binding DNA and cutting DNA or altering gene expression at specific DNA sites. However, the tools cannot repair or necessarily reverse their effects. If the genome editing tool cuts the DNA, it must rely on the cell's machinery to repair the break, otherwise the cell will not survive. If the genome editing tool modifies the DNA, the degree to which the effect can be reversed depends on the mechanisms by which the tools act. In addition, CRISPR-based tools have been developed to push desired traits through a population of organisms; this tool is referred to as a gene drive. Gene drives are feasible in organisms that have short generation times, such as small mammals (e.g., mice) and insects (e.g., mosquitoes). These tools are advancing at an extremely fast pace, adding new capabilities rapidly.

These tools have raised considerable interest within the life science community to enable precise editing of human, animal, and plant genomes to correct deleterious or disease-causing sequences in the DNA. At the same time, members of the security community have expressed concern that these tools can enable adversaries to insert harmful genes into people, animals, or plants; enhance advantageous traits within their population; create harmful pathogens; or drive harmful traits through a population. The security community has speculated that genome editing tools lower the technical barriers to modifying organisms and microbes. However, genome editing tools are becoming ubiquitous throughout the world, especially as their application to addressing social issues gains support. For example, the promise of correcting a disease-causing mutation in people or introducing disease-preventing traits in people serves as a primary driver for advancement and application of genome editing tools in people, both in somatic cells and in embryos. Scientists in China, the U.S., and Europe have published research on precise modification of human and animal somatic cells [41] and viable human embryos [30]. In 2018, a Chinese researcher modified human embryos inserting the CCR5 deletion known to protect people from R5 HIV strains and allowing the modified embryos to be carried to term [25]. Furthermore, genome editing tools are being used in agriculture to make crops more resistant to pests [39] and livestock that are safer for humans to handle [4]. In addition to these applications, genome editing tools are improving research by enhancing genetic screening efforts that link specific genomic sequences to traits, enabling epigenetic screening to identify the molecular determinants of traits that are not encoded in the genome, and improving capabilities for temporarily or permanently modifying microbes and cells. Furthermore, genome editing tools now are being applied to pest control, such as the creation of sterile mosquitoes and mice through the development of gene drives in these organisms [3, 31, 33].

Whether and to what degree these capabilities present a security risk remains a question. In 2016, when the U.S. Office of the Director of National Intelligence called gene editing a weapon of mass destruction, the White House Office of Science and Technology Policy initiated a process for reviewing the security risks of genome editing tools and associated biotechnologies. The primary concerns expressed during this timeframe focused on the creation or modification of harmful microbes. However, the technical advantages afforded by genome editing tools over traditional genetic engineering approaches for microbes are limited to specific microbes. For example, most bacteria are unable to repair double-strand breaks because they do not have the same repair machinery as more complex organisms. To address this technical issue, scientists have created a suite of products that supply the missing repair enzyme along with the CRISPR-based tool. Another example is modification of viruses that are difficult to engineer, such as Epstein-Barr Virus [40]. However, existing methods for genetically modifying or synthesizing other viruses are as effective as genome editing approaches. For example, in 2005 scientists used reverse genetics to resurrect the 1918 influenza virus [35], which led to many findings about how the virus caused disease in humans. Three years before that, scientists at State University of New York at Stoney Brook synthesized infectious poliovirus [5]. In the 2010s, the J. Craig Venter Institute chemically synthesized full length and minimal bacterial genomes, both of which resulted in live bacteria, after much trial-and-error [13, 18]. In 2018, scientists in Canada created horsepox virus from published sequence (WHO 2017). None of these efforts involved the use of genome editing tools. Therefore, the potential security risks unique to genome editing may be associated with modifying more complex organisms and changing the distribution of naturally-occurring traits rather than creation of harmful pathogens. However, the capability to edit more complex organisms rests primarily with well-resourced, sophisticated, and skilled organizations, indicating that actors with access to a well-equipped laboratory, biological materials, and funds may have the requisite capabilities to use genome editing tools to harm adversaries. In addition, these organization likely have access to the information needed to design and develop modified organisms using precise genome editing.

Numerous genome editing products and services are offered commercially, including plasmid DNA with the genome editing tools, genome editing tools that are coupled to gene therapy vectors (e.g., adenovirus or adeno-associated virus), custom viruses, services to assist in the design of guide RNA for CRISPR-based systems, edited cells (e.g., seeds and cattle sperm), and libraries of guide RNA. The number and diversity of products and services are growing at a significant rate as commercial and academic institutions develop new tools. Despite this robust commercial market, the ability to use the tools is limited by the capabilities for gene delivery (in more complex organisms) and the available (and reproducible) knowledge about genetic or epigenetic determinants of traits. As associated information and technologies are developed, the potential risk of genome editing may change. However, at this time, several challenges limit the predictability of genome editing:

- Detailed understanding about the functions of specific sequences in heterogenic, diverse populations is limited and often, irreproducible.
- The effects of specific changes in different genetic backgrounds is not well known in diverse populations.
- Observed traits can be derived from the multiple genes, regulated at the genomic level or after the messenger RNA is made, and/or after a protein is made. This level of complexity increases uncertainty in genome-wide association studies, which do not necessarily account for epigenetic patterns or regulation.

- Current computer-based approaches for designing specific genetic changes depend on the input of reliable, complete, and defensible genomic, trait association, and epigenomic data, any of which are extremely limited for many organisms.

Furthermore, consequences of edited organisms are not well-understood, which significantly limits any qualitative or quantitative risk analyses. The effects of genome editing cause on- and off-target effects (i.e., mutation at the intended location in the DNA and mutation at unintended locations in the DNA, respectively). However, the ability to assess health and physiological consequences of genome editing depends greatly on the organism and availability of analytic methods, which include computational or statistical models, biochemical or molecular biology assays, or observation of traits. Distinguishing between mutations created through the editing process and previously undocumented natural variants is challenging and suggests the need for more controlled, thorough scientific studies to ensure that any assignment of cause-and-effect of specific sequences or epigenetic patterns are a result of genome editing rather than nature. Furthermore, predicting the potential effects of genome editing is difficult despite some modifications having been induced reliably in agriculturally important organisms. Modification of many animals, plants, and humans is extremely difficult because of the high degree of genomic diversity, limited information about trait associations, imperfect data, and suboptimal gene delivery tools.

Assessing environmental consequences of edited organisms that may be released into the wild (e.g., sterile mosquitoes or microbes created to degrade hazardous chemicals) is significantly limited by the lack of knowledge about ecological effects of introducing a laboratory-modified organism into the wild. Relevant factors needing to be considered include fitness of the edited organism, persistence of the modification (i.e., the natural pressures for maintaining, changing, or compensating for the mutation), adaptation of the edited organism, behavioral patterns caused by laboratory procedures (e.g., flight patterns of mosquitoes in the laboratory setting compared to natural mosquito flight patterns), and the baseline of natural organisms and their genomic and epigenetic patterns. Over the past several years, efforts to sequence microbes, insects, and plants in different environments have been undertaken by academic, industrial, and DIY biology community with government, and personal or crowdsourced funds. These efforts may help provide a snapshot of organisms and their variants that exist in nature. However, these sampling and sequencing efforts must be conducted longitudinally to ensure that natural mutation is captured. Otherwise, the same challenges in distinguishing a mutation caused by genome editing from natural variation persist and questions about genetic (or epigenetic) determinants of observed traits exist.

Although genome editing tools are available through academic and commercial sources, they still require knowledge, skills, specialized experimental technologies, and financial resources to design and use these tools in various organisms, and may have many unintended (and as yet, undocumented) consequences.

4.4 Big Data in the Life Sciences

Big data analytics is defined by industry leaders as the analysis of large sets of data (volume), from many different sources (variety), with variable degrees of certainty and completeness (veracity), and quickly (velocity) [22] Companies, such as IBM, have been an intellectual and product leader in the big data field, being the first technology company to engage with the scientific community through its Watson program [21]. Various applications of big data analytics followed shortly thereafter, with efforts in precision medicine, living healthcare systems, precision agriculture, and analysis of ecosystems and environments. These fields leverage the advances in data science and engineering to integrate, analyze, and visualize all kinds of data, including text, image, and audio. For example, advances in natural language processing have enabled technology companies, healthcare systems, and researchers to analyze large amounts of published literature, news and popular press articles, data in databases, and social media posts written in many languages. Similarly, image analysis software enables scientists to analyze photos, videos, and other images. New methodologies for artificial intelligence and machine learning uses training and verification data sets to teach computational and mathematical algorithms to analyze data based on scientific knowledge. In addition, the metadata associated with these data files (e.g., date, time, and GIS location) can be included in analyses, enhancing the analytic results.

During the past decade, the U.S. government has established several significant programs in big data: (1) the Big Data 2 Knowledge (BD2K) initiative, which is a NIH effort; (2) the BRIAN Initiative, which is funded by the NIH and U.S. Department of Defense; (3) Precision Medicine Initiative, which involves several U.S. agencies; (4) NEON, which is an effort supported by the National Science Foundation (NSF) to understand the environment better; (5) the microbiome project, which for humans is supported by the NIH and for ecosystems is supported by the NSF and U.S. Department of Energy; (6) biosurveillance, which is funded by several U.S. agencies, international organizations, and companies; (7) the Global Virome Project, which is supported in part by the U.S. Agency for International Development; [41] and (8) precision agriculture, which is supported largely by the agricultural sector. All of these efforts build on established research efforts in systems biology, in which cellular and organismal processes and functions are examined as systems. Although initial efforts used data from publications, more recent efforts have leveraged the vast amounts of data being produced from various sources, including -omics data (e.g., genomics (i.e., the study of an organism or cell's genome), transcriptomics (i.e., the study of gene expression patterns in a cell or organisms), and metabolomics (i.e., the study of metabolites and biosynthetic pathways in a cell or organism)). The BD2K, BRAIN Initiative, and Precision Medicine Initiatives involve the integration of various -omics data. Within the last eight years, a new -omics approach has emerged: the exposome. The exposome integrates various -omics data to examine the changes that occur in a cell or an organism after exposure to various stimuli or hazards.

The advances in data generation and analysis have raised considerable interest among a variety of stakeholders, including academia, industry, and government researchers, governmental decision-makers, private organizations, and the health and agriculture industries. However, the way in which big data in the life sciences emerged coupled with the speed with which it has evolved and been applied has raised concern among security experts about the potential for the data and analytic processes to be stolen or diverted for malicious purposes. This concern was raised in 2014 when AAAS, FBI, and UNICRI published their study on the national and transnational security implications of big data in the life sciences [32]. This study highlighted the potential vulnerabilities of big data to cyberattack, wherein an individual or group could access databases, stealing the data for some unknown use, manipulating data, holding data ransom, or flooding data analysis platforms to either shut down the system or dilute signals. As the study was being conducted, several healthcare facilities and insurance companies were attacked, with only healthcare data being stolen for purposes that are not evident. Shortly thereafter, private companies and the U.S. Office of Personnel Management records were attacked. The prevalence of these events further supported the real risk that life science databases are vulnerable to cyberattacks. The purpose of these attacks has not been described, but two possibilities are that the information contained within could enable entities to gain competitive advantage in the pharmaceutical market or to identify people through their genetic sequences. Although actors of all types could attack computer systems, the size and complexity of the data suggests that sophisticated, well-resourced actors have the data storage and analytic capabilities to analyze the stolen information.

Theft of information is not the only approach used by adversaries. Concerns about funding and technologies or skills transfer of have been raised by law enforcement. Over the past five years, the FBI has alerted industry, academic, and international stakeholders about tactics that adversaries may use to steal valuable information, technologies, or skills. Among the approaches used by foreign governments, adversarial groups, and competitors is the acquisition of information, technologies, and/or skills through funding research of interest or acquiring, investing in, or merging with companies. These acquisition efforts may be direct as in the example of the commercial acquisition of a gene sequencing company, Complete Genomics, after which the parent company takes all of the modern sequencing equipment and sends its personnel for training at the acquired company. Alternatively, the acquisition efforts may be indirect, as in the purchase of agricultural land on which farmers may use commercially available seeds to grow crops. In this example, the adversary may be interested in information about the seeds (e.g., its genomic sequence or any modifications that make it resistant to pests), ecology of the soil, yield efficiencies, or other data of possible interest. In these scenarios, the information, tools, and skills are valuable to an organization, including those that may be financed by or otherwise linked to an adversarial government, which could present national security risks. Furthermore, the use of funding of research to obtain data raises several questions, especially considering that about 50% of U.S. research is funded by the U.S. government. The trends towards cross-over venture capital and crowdsourcing of research, where the funder stands to gain financially if their investment pays off, may be a sign that even seemingly legitimate research may have ulterior purposes.

These concerns are most closely associated with economic, commercial, and agricultural security. However, big data, particularly those associated with personalized medicine, may present national security risks to human health. Although most of the investment currently is in defining molecular determinants of traits and disease states, the information gained through these efforts may be used by adversaries to specifically enhance desired traits in their populations or to target certain populations. The feasibility of these applications still seems years (maybe decades) away. However, recent efforts to improve genetic association studies may be exploitation resulting in such malicious efforts. Therefore, the FBI and others have begun engaging scientific and technical stakeholders in safeguarding their data to ensure their use in legitimate, beneficial studies only.

In general, adversaries of all types can carry out cyberattacks on databases, data systems, and computer networks. However, only those actors that have the intent, data and analytic capabilities present higher risks for malicious use of information, technologies, and skills used in big data efforts. Furthermore, the knowledge and skill needed to analyze big data in the life sciences may be sophisticated and exploited by well-trained, well-resourced entities.

5 Summary

For nearly 20 years, the scientific and security communities have struggled with defining and reviewing life sciences research that may have dual use potential. These challenges become increasingly more difficult as new biotechnologies are created, refined, and applied to different scientific and societal problems. To date, the primary approaches for assessing the dual use potential of life sciences research and biotechnologies have been: (1) to examine whether the technology or scientific research could be used to make previously non-existent pathogens, convert harmless microbes to harmful pathogens, and/or generate pathogens more easily and with fewer resources or skills; and (2) to determine whether a research matches any of the seven experiments of concern originally listed by the National Research Council report, *Biotechnology in an Age of Terrorism*, and more recently, the U.S. government policies on dual use research of concern. However, these approaches have proven to be limiting, especially when examining potential dual use implications of biotechnologies that have primary applications in life science research that does not involve pathogens or microbes. For example, the national security concerns about genome editing and big data in the life sciences are based on capabilities developed and applied to non-pathogen research, despite some policymakers focusing heavily on their narrow application to pathogen research. This limitation results in biased risk assessment, which has led to policy and regulatory suggestions that could have significant unintended consequences to research, especially if they affect a larger, more diverse array of studies than those considered.

Multidisciplinary research will continue to generate new biotechnologies, scientific approaches, and knowledge. Further advancement and application of new technologies and scientific information will be driven by societal need, whether for agriculture, health, environmental preservation, energy, or defense. Technologies may be developed and/or used by citizen and professional scientists and engineers, and may be supported by private, crowdsourced, or foreign government funds. The consequences of theft or diversion of beneficial or legitimate biotechnologies could involve economic, commercial, and societal harms rather than human health harms, which is the focus of most current discussions on dual use life sciences research. Furthermore, adversaries (insiders, individuals, groups, or foreign governments) will exploit biotechnologies if they believe the tools can contribute to their strategic or tactical objectives, and they have the requisite resources, knowledge, access, and skill needed to obtain and apply the information and/or technologies.

Looking forward, the scientific and security communities may need to consider a new approach to evaluating the potential national security risks of emerging biotechnologies. The security community could define the national security risks about which they are most concerned, such as loss of trade for a major agricultural commodity or severe societal disruption. Currently, only human health consequences are defined, but those effects primarily are restricted to effects caused by pathogens. Broadening consequences of interest beyond only human health outcomes enables a more complete assessment of risk and benefit of biotechnologies. The next step would involve security experts focusing their concerns on the specific types of events that might lead to those defined consequences. This step would result in a list of events of particular concern, providing the lens through which scientists and security experts could evaluate the potential for harmful applications of biological sciences and technology. This concept is not dissimilar to the recommendations of the National Research Council. However, implementation of those recommendations seems to have veered in a different direction, specifically one that focuses more on a select set of pathogens with which research is conducted.

Analyses could be enhanced through open communication and dialogue with other stakeholders, ensuring that an evaluation of risk is based on actual scientific capabilities and trends, on available commercial products and services, and on actual adversary capabilities, resources, knowledge, and skill. Ideally, this process would allow for more thorough assessment of potential risks, increasing the potential that high and moderate security risks can be addressed early and reducing the possibility that legitimate and/or beneficial research would stop or slow. This approach is not predicated on examining one biotechnology advance or capability at a time. Rather, it focuses on capabilities that might lead to concerning events and consequences, which provides a more systematic way of identifying actual vulnerabilities and developing or leveraging measures for reducing or eliminating those vulnerabilities.

References

1. American Society of Microbiology. Statement of Journal Editors and Authors Group on Scientific Publishing and Security. (2003). https://www.asm.org/index.php/position-statements-and-testimony?id=2453
2. American Society of Microbiology (2003) Statement of Journal Editors and Authors Group on Scientific Publishing and Security. http://www.asm.org/index.php/position-statements-and-testimony?id=2453
3. Brown, K. (2017). New Zealand could use gene editing To Kill Off Its Cutest Preditor. https://www.gizmodo.com.au/2017/04/new-zealand-could-use-gene-editing-to-kill-off-its-cutest-predator/
4. Carlson, D. F., Lancto, C. A., Zang, B., Kim, E. S., Walton, M., Oldeschulte, D., . . . Fahrenkrug, S. C. (2016). Production of hornless dairy cattle from genome-edited cell lines. *Nat Biotechnol, 34*(5), 479-481. doi: https://doi.org/10.1038/nbt.3560
5. Cello J, Paul AV, Wimmer E (2002) Chemical synthesis of poliovirus cDNA: generation of infectious virus in the absence of natural template. Science 297(5583):1016–1018. https://doi.org/10.1126/science.1072266
6. Considerations in the regulation of biological research (1978) Univ PA Law Rev 126(6):1420–1446
7. Defense Advanced Research Projects Agency (2016) Defense Advanced Research Projects Agency: Biological Technologies Office. Retrieved November 4, 2017. http://www.darpa.mil/about-us/offices/bto?PP=3
8. Defense Advanced Research Projects Agency (2017). Building the Safe Genes Toolkit. http://www.darpa.mil/news-events/2017-07-19
9. Defense Advanced Research Projects Agency (2014) DARPA Launches Biological Technologies Office. http://www.darpa.mil/news-events/2014-04-01
10. Defense Advanced Research Projects Agency. (2014). DARPA Launches Biological Technologies Office. https://www.darpa.mil/news-events/2014-04-01
11. Defense Advanced Research Projects Agency. (2016). Defense Advanced Research Projects Agency: Biological Technologies Office. https://www.darpa.mil/about-us/offices/bto?PP=3
12. Defense Advanced Research Projects Agency (2017). Building the Safe Genes Toolkit. https://www.darpa.mil/news-events/2017-07-19
13. Gibson, D. G., Glass, J. I., Lartigue, C., Noskov, V. N., Chuang, R. Y., Algire, M. A., . . . Venter, J. C. (2010). Creation of a bacterial cell controlled by a chemically synthesized genome. *Science, 329*(5987), 52-56. doi: https://doi.org/10.1126/science.1190719
14. Global Virome Project. (n.d.). *Global Virome Project.* https://www.globalviromeproject.org/
15. Global Virome Project (2017). http://www.globalviromeproject.org/
16. Gryphon Scientific. (2018) Training Materials for Practical Implementation of Laboratory Biosafety, Biosecurity, and Biorisk Management. Accessible at https://www.gryphonscientific.com/resources/training-materials-for-practical-implementation-of-laboratory-biosafety-biosecurity-and-biorisk-management/.
17. Gryphon Scientific (2018) Training Materials for Practical Implementation of Laboratory Biosafety, Biosecurity, and Biorisk Management. Accessed on August 17, 2020 . Accessible at https://www.gryphonscientific.com/resources/training-materials-for-practical-implementation-of-laboratory-biosafety-biosecurity-and-biorisk-management/
18. Hutchison, C. A., 3rd, Chuang, R. Y., Noskov, V. N., Assad-Garcia, N., Deerinck, T. J., Ellisman, M. H., . . . Venter, J. C. (2016). Design and synthesis of a minimal bacterial genome. *Science, 351*(6280), aad6253. doi: https://doi.org/10.1126/science.aad6253
19. IBM. (n.d.) The Four V's of Big Data. https://www.ibmbigdatahub.com/infographic/four-vs-big-data
20. IBM Research. (n.d.). *Computational Biology.* https://researcher.watson.ibm.com/researcher/view_group.php?id=137
21. IBM (2017) Research Computational Biology. http://researcher.watson.ibm.com/researcher/view_group.php?id=137

22. IBM (2017) The Four V's of Big Data. http://www.ibmbigdatahub.com/infographic/four-vs-big-data

23. Kupferschmidt, K. (2017). How Canadian Researchers Reconstituted an Extinct Poxvirus for $100,000 Using Mail-Order DNA. *Science.*

24. Marburger, L., Schaub, K., Celeste, E., Geers, G., AlHmoud, N., benFradj, O., . . . Berger, K. (Producer). (2014, November 4, 2017). International Engagement: Secure Science, Technology, and Research - BMENA Case Studies.

25. Marchione, Marilynn. (2018) Chinese Researcher Claims First Gene-Edited Babies. Associated Press. Accessible at: https://apnews.com/4997bb7aa36c45449b488e19ac83e86d.

26. National Research Council (1982) Scientific Communication and National Security XE "National Security" . National Research Council, Washington, DC

27. National Research Council (2004) Biotechnology Research in an Age of Terrorism. National Academy Press, Washington, DC

28. National Research Council (2011) Research in the Life Sciences with Dual Use Potential: An International Facility Development Project on Education About the Responsible Conduct of Science. National Academies Press, Washington, DC

29. Nuclear Threat Initiative. (2012). Between Publishing and Perishing? H5N1 Research Unleashes Unprecedented Dual-Use Research Controversy. https://www.nti.org/analysis/art icles/between-publishing-and-perishing-h5n1-research-unleashes-unprecedented-dual-use-research-controversy/

30. Park, A. (2017). Researchers are Now Editing Genome of Human Embryos. *Time.*

31. Regalado, A. (2016). Bill Gates doubles his bet on wiping out mosquitoes with gene editing. *MIT Technology Review.*

32. Roderick, J., Bashour, N., Kang, D., Lee, B., Kim, E., Therrien, J., . . . Berger, K. (2014). National and Transnational Security Implecations of Big Data in the Life Sciences. Washington, DC: American Association for the Advancement of Science, Federal Bureau of Investigation, United National Institute for Interregional Crime and Justice Research Institute.

33. Thompson, A. (2017). Scientists Want to Use a 'Gene Drive' to Wipe Out Invasive Mice. *Popular Mechanics.* https://www.popularmechanics.com/science/animals/a25203/gene-drive-wipe-out-invasive-mice/

34. Tucker J (2012) Innovation, Dual Use, and Security: Managing the Risks of Emerging Biological and Chemical Technologies. MIT Press, Boston, MA

35. Tumpey, T. M., Basler, C. F., Aguilar, P. V., Zeng, H., Solorzano, A., Swayne, D. E., . . . Garcia-Sastre, A. (2005). Characterization of the reconstructed 1918 Spanish influenza pandemic virus. *Science, 310*(5745), 77-80. doi: https://doi.org/10.1126/science.1119392

36. Universtiy of Penn Law Review (1978) Considerations in the regulation of biological research. Univ PA Law Rev 126(6):1420–1446

37. Wegrzyn, R. (2017). SafeGenes. https://www.darpa.mil/program/safe-genes

38. World Health Organization. (2017). WHO Advisory Committee on Variola Virus Research: Report of the Eighteenth Meeting.

39. Xiong JS, Ding J, Li Y (2015) Genome-editing technologies and their potential application in horticultural crop breeding. Hortic Res 2:15019. https://doi.org/10.1038/hortres.2015.19

40. Yuen, K. S., Chan, C. P., Wong, N. H., Ho, C. H., Ho, T. H., Lei, T., . . . Jin, D. Y. (2015). CRISPR/Cas9-mediated genome editing of Epstein-Barr virus in human cells. *J Gen Virol, 96*(Pt 3), 626-636. doi: https://doi.org/10.1099/jgv.0.000012

41. Zafra MP, Dow LE (2016) Somatic Genome Editing XE "Genome Editing" Goes Viral. Trends Mol Med 22(10):831–833. https://doi.org/10.1016/j.molmed.2016.08.004

Index

© Springer Nature Switzerland AG 2021

R. N. Burnette (ed.), *Applied Biosecurity: Global Health, Biodefense, and Developing Technologies*, Advanced Sciences and Technologies for Security Applications, https://doi.org/10.1007/978-3-030-69464-7